EASY CHEMISTRY

EIGHTH EDITION

LOUIS CASIMIR

SCIENCE INSTRUCTOR

ST. AUGUSTIN COLLEGE SCIENCE INSTRUCTOR

NORTHEASTERN ILLINOIS UNIVERSITY

All inquiries should be addressed to:

Book Domain LLC.
543 E Louise Dr Phoenix, Az 85050

Ordering Information:

Amount Deals. Special rebates are accessible on the amount bought by corporations, associations, and others. For points of interest, contact the distributor at the address above.

Printed in the United States of America.

ISBN-13 Paperback 978-1-967903-65-8
eBook 978-1-967903-64-1

ACKNOWLEDGEMENTS

I would like to thank the following chemists and educators for reviewing Easy Chemistry.

Dr. Regnal Jones
Executive Director of CAHMCP
Illinois Institute of Technology
Chicago, Illinois

Mr. Tom Bradley
Chemistry Department CAHMCP
Illinois Institute of Technology

Mr. Thomas Bearden, Principal
St. Gregory High School
Chicago, Illinois

Ms. Tamu Hunter
Research Chemist
University of Chicago Chicago, Illinois

Mrs. Lucy McCorkle
Public Schools
Washington, D.C.

Dr. Michael Okorafor
Assistant Professor of Chemical Engineering
University of Detroit
Detroit, Michigan

Sister Mary Cormille
Former Principal and Science Professor
St. Gregory High School
Chicago, Illinois

Mr. Jerry Kiziel, Chemistry Professor
St. Frances DeSales
Chicago, Illinois

Dr. Lance Williams
University of Chicago Testing Programs
Biology and Chemistry
Chicago, Illinois

EASY CHEMISTRY and **UNIQUE CHEMICAL LABS,** both
by Louis Casimir

TABLE OF ATOMIC WEIGHTS

(Based on Carbon-12)

Name	Symbol	Atomic Number	Atomic Weight	Name	Symbol	Atomic Number	Atomic Weight
Actinium	Ac	89	(227.0)	Mendelevium	Md	101	(258)
Aluminum	Al	13	26.98154	Mercury	Hg	80	200.5
Americium	Am	95	(243)	Molybdenum	Mo	42	95.94
Antimony	Sb	51	121.75	Neodymium	Nd	60	144.24
Argon	Ar	18	39.948	Neon	Ne	10	20.179
Arsenic	As	33	74.9216	Neptunium	Np	93	237.0482
Astatine	At	85	(210)	Nickel	Ni	28	58.70
Barium	Ba	56	137.33	Niobium	Nb	41	92.9064
Berkelium	Bk	97	(247)	Nitrogen	N	7	14.0067
Beryllium	Be	4	9.01218	Nobelium	No	102	(255)
Bismuth	Bi	83	208.9804	Osmium	Os	76	190.2
Boron	B	5	10.81	Oxygen	O	8	15.9994
Bromine	Br	35	79.904	Palladium	Pd	46	106.4
Cadmium	Cd	48	112.41	Phosphorus	P	15	30.97976
Calcium	Ca	20	40.08	Platinum	Pt	78	195.09
Californium	Cf	98	(251)	Plutonium	Pu	94	(244)
Carbon	C	6	12.011	Polonium	Po	84	(210)
Cerium	Ce	58	140.12	Potassium	K	19	39.0983
Cesium	Cs	55	132.9054	Praseodymium	Pr	59	140.9077
Chlorine	Cl	17	35.453	Promethium	Pm	61	(147)
Chromium	Cr	24	51.996	Protactinium	Pa	91	231.0359
Cobalt	Co	27	58.9332	Radium	Ra	88	226.0254
Copper	Cu	29	63.546	Radon	Rn	86	(222)
Curium	Cm	96	(247)	Rhenium	Re	75	186.207
Dysprosium	Dy	66	162.50	Rhodium	Rh	45	102.9055
Einsteinium	Es	99	(254)	Rubidium	Rb	37	85.4678

Erbium	Er	68	167.26	Ruthenium	Ru	44	101.07
Europium	Eu	63	151.96	Samarium	Sm	62	150.4
Fermium	Fm	100	(257)	Scandium	Sc	21	44.9559
Fluorine	F	9	18.998403	Selenium	Se	34	78.96
Francium	Fr	87	(223)	Silicon	Si	14	28.0855
Gadolinium	Gd	64	157.25	Silver	Ag	47	107.868
Gallium	Ga	31	69.72	Sodium	Na	11	22.98977
Germanium	Ge	32	72.59	Strontium	Sr	38	87.62
Gold	Au	79	196.9665	Sulfur	S	16	32.06
Hafnium ±	Hf	72	178.49	Tantalum	Ta	73	180.9479
Hahnium	Ha	105	(260)	Technetium	Tc	43	98.9062
Helium	He	2	4.00260	Tellurium	Te	52	127.60
Holmium	Ho	67	164.9304	Terbium	Tb	65	158.9254
Hydrogen	H	1	1.0079	Thallium	Tl	81	204.37
Indium	In	49	114.82	Thorium	Th	90	232.0381
Iodine	I	53	126.9045	Thulium	Tm	69	168.9342
Iridium	Ir	77	192.22	Tin	Sn	50	118.69
Iron	Fe	26	55.847	Titanium	Ti	22	47.90
Krypton	Kr	36	83.80	Tungsten	W	74	183.85
Kurchatovium±	Ku	104	(259)	Uranium	U	92	238.029
Lanthanum	La	57	138.9055	Vanadium	V	23	50.9415
Lawrencium	Lr	103	(260)	Xenon	Xe	54	131.30
Lead	Pb	82	207.2	Ytterbium	Yb	70	173.04
Lithium	Li	3	6.941	Yttrium	Y	39	88.9059
Lutetium	Lu	71	174.967	Zinc	Zn	30	68.38
Magnesium	Mg	12	24.305	Zirconium	Zr	40	91.22

The above values apply to elements as they exist in materials of terrestrial origin and to certain artificial elements. They are reliable to ± 1 in the last digit or ± 3 if that last digit is below the line. Weights given in parentheses are approximate only and have not been officially accepted.

PREFACE

The purpose of **EASY CHEMISTRY** is two-fold. The first is to introduce a few basic concepts of chemistry. The second is to make chemistry a pleasant, understandable and applicable science. Moreover, we want to avoid memorization of formulas and encourage logical thinking and reasoning.

EASY CHEMISTRY is dedicated to my parents Samuel and Elizabeth Casimir.

ABOUT THE AUTHOR

Louis Casimir has taught chemistry at both the high school and college level for more than 16 years. His first chemistry text, EASY CHEMISTRY, has helped to motivate thousands of students. Professor Casimir emphasizes practical applications of chemistry and logical reasoning.

SPECIAL THANKS

Secretarial Staff
Gladys O. Bustamante Tracy Wilson Gregory Karen Cooper
Brow Word Processing
Mayra R. Quinones
Sylvia James
Maria Christina Garcia
Reality Anderson
Tessa Jones

CONTENTS

LECTURE TOPICS

EASY CHEMISTRY OBJECTIVES AND LEARNING GOALS

Learning Goals

Students Should Be Able To:

1. Convert units of mass, length, and volume within the metric system using the factor-unit method.
2. Convert from a metric unit to the corresponding English unit using the factor-unit method.
3. Calculate the density; mass, or volume of an object when you are given the other two.
4. Find the number of significant figures in a measurement
5. Do calculations using the rules for significant figures
6. Write numbers in scientific notation.
7. Convert temperatures from the Celsius to the Kelvin scale, and vice versa.
8. Explain the Law of Conservation of Mass and Energy.
9. Explain the difference between physical and chemical properties.
10. describe the difference between an atom and a molecule.
11. Explain the terms atomic mass, atomic weight, molecular weight, gram-molecular weight, mole, and Avogadro's number.
12. Figure out the number of moles in a sample of an element when you are given the weight of the sample.
13. Figure out the weight, in grams, of a sample when you are given the number of moles.
14. Figure out the empirical formula for a compound when you are given its percentage composition.
15. Determine the molecular weight of a compound when you are given the formula for the molecule.
16. Determine the molecular formula of a compound from its empirical formula and molecular weight.
17. Describe Daltons atomic theory: Thomsons model of the atom, and Rutherford's model of the atom.
18. Give the charge and mass of the electron, proton, and neutron.
19. Explain what isotopes are, and give examples.
20. Use the standard notation for isotopes.
21. Describe the Bohr model of the atom.
22. Explain shells and energy levels.
23. Write electron configuration for element 1 through 20.
24. State what is meant by the octet rule.
25. Distinguish between a period and a group in the periodic table.

26. State the difference between an A-group (representative) element and a B-group (transition) element.
27. Predict trends for properties such as atomic radius, ionization potential, and electron affinity.

After Easy Chemistry: Students Should Be Able To:

28. Write the electron configurations for various elements, using the filling-order diagram.
29. Define and discuss atomic orbitals.
30. Draw shapes for the s and p orbitals.
31. Write Lewis (electron-dot) structure for the A-group elements.
32. Define and give an example of a covalent bond.
33. Name the diatomic elements from memory.
34. Write electron-dot structures for various covalent compounds.
35. Define and give an example of an ionic bond.
36. Write the chemical formula for some compounds when you are given their chemical names.
37. Find the oxidation numbers of less familiar elements when you are given chemical formulas of compounds containing these elements.
38. Write the chemical names of some compounds when you are given the formula for the compounds'.
39. Distinguish between a polar and a nonpolar bond.
40. Write formula equations from word equations, and vice versa.
41. Balance formula equations.
42. Recognize and give examples of some general types of chemical reactions (combination, decomposition, single-replacement, double-replacement, and combustion.

CHAPTER 1

Math Review and Metric Conversions

Objectives

To be able to:

- review basic math concepts
- to be able to convert from metric to English
- to study three states of matter
- to study temperature scales and conversions

Keywords:

- metric
- meter
- matter
- solid
- mass
- density
- volume
- factoring
- exponents
- kilo
- centi
- milli
- deci

CHAPTER 1

MEASUREMENT AND METRIC TOPICS

Measurement is essential to master chemistry. Since chemistry is the study of matter and its changes in composition, we must learn how to measure some of these changes.

BASIC MATH REVIEW

Decimals: Based on power of ten.
10 x 1 - 10 or 10.00
10 x 4 40 or 40.00
1/100 = 0.01

Fractions: Denominator = 100% Numerator = Part of 100%

$$\frac{1}{2} = .5 \; or \; 2\overline{)\sqrt{10}}^{\,0.5}$$

$$\frac{2}{2} = 100\% \qquad \frac{4}{5} = \frac{0.8}{5\sqrt{4.0}} = 0.8 \; or \; 80\%$$

Rounding off

If greater than .5, use next whole number. Example: 1.58 = 1.6

Multiplication

6 x 3.8 = 22.8

Use of Decimals

Decimals are based on power of ten.

1. Moving decimal point to the right increases value.
2. Moving decimal point to the left decreases value.

Example of decimal point to the right:
30.00 30.0 = 300 300.0 = 3,000

Use of Decimals

Example of decimal point to the left:

5,000 =- 500.0 = 500 50.0 = 50 5.0=5

Percent and Decimals

To convert decimals to percentages (%), we move *two* places to the *right*:

.6 = 0.60 = 60%

To convert percentages (%) to decimals, we move *two* places to the *left*:

80% = 80 or 0.80

Fractions

Fractions and decimals can be interchangeable because they can be 100% or part of the whole.

Example of Fractions

3/3 = 100% 1/3 = 33%

Note: When numerator or top number is the same as denominator or bottom number, we get 100%. Likewise, when numerator is less than denominator, we have less than 100%. If numerator is greater than denominator, we have more than 100%.

Example:

1/8 = 0.125 or 12.5% 3/9 = 0.333 or 33.3% ¼ = 0.25 or 0.25

27/9 = 3 9/9 = 1

$\frac{40}{400} = 0.01$ $\frac{400}{1000} = 0.04$ $\frac{4000}{1000} = 4$ $\frac{4}{1000} = 0.004$

Significant Figures and Rounding Off

1. Zeroes to the left of the decimal point are insignificant.

   0.8 = 8 00.8 = 8 or 0.8

2. Any numbers to the right are significant.

   Example: 13.678 has 5 significant figures

   1.067 has 4 significant figures

Review basic math book for more details.

Exponents

Chemists and other scientists often use small and large numbers. To simplify operations, they use *exponential* or *scientific notation.*

Examples:

4	$= 4.0 \times 10^0$
40	$= 4.0 \times 10^1$
400	$= 4.0 \times 10^2$
4,000	$= 4.0 \times 10^3$
40,000	$= 4.0 \times 10^4$
400,000	$= 4.0 \times 10^5$
4,000,000	$= 4.0 \times 10^6$ (4 million)
40,000,000	$= 4.0 \times 10^7$
400,000,000	$= 4.0 \times 10^8$
4,000,000,000	$= 4.0 \times 10^9$ (4 billion)

Note: the zeros equal the exponents (4.0×10^2 has two zeros).

Working With Exponents

- **Multiplication**

  $(3.0 \times 10^{-5}) \times (2.0 \times 10^{8}) = 6.0 \times 10^{3}$

  Simplified Whole Numbers: $3 \times 2 = 6$

  Exponents: $-5 + 8 = 3$

  The sum $= 6.0 \times 10^{3}$

  When multiplying exponents, add the exponents.

  $6.0 \times 10^{3} \times 1.5 \times 10^{2} = 9.0 \times 10^{5}$ or $6{,}000 \times 150 = 900{,}000$

- **Dividing Exponents**

  When dividing exponents, we subtract.

  $$\frac{8.0\ x\ 10^{-6}}{2.0\ x\ 10^{-5}} = 4.0\ x\ 10^{-11}$$

Dividing Exponents

Another examples of division:

$$\frac{(2.0\ x\ 10^{4})\ x\ (6\ x\ 10^{-6})}{2.0\ x\ 10^{-7}} = \frac{1.2\ x\ 10^{-1}}{2\ x\ 10^{-7}} = 6.0\ x\ 10^{5}$$

METRIC SYSTEM

In our Easy Chemistry text we will study a few common of the Metric System. Metric Conversions

A. Convert **40 cm** to meters. Since *centi* is 1/100 we only change the numerator:

Solution: $\frac{40\cancel{cm}\ x\ 1m}{100\cancel{cm}} = 0.4meters$

B. Convert 125 cm to meters (m).

Solution : $Centi = \frac{1}{100}, so\ 125\ \cancel{cm}\ x \frac{1m}{100\cancel{cm}} = 1.25meters$

C. Convert 5 cm to meters.

Solution: $5\cancel{cm}\ x \frac{1m}{100\cancel{cm}} = 0.05meters$

D. How many ml (milliliters) in a 2 liter bottle of cola?

Solution: $1{,}000\ mL = 1 liter, so\ 2\not{L}\ x \frac{1{,}0000 mL}{1\not{Liter}} = 2{,}000 mL$

E. A lab student has 300 mL of flammable 0_2 gas. How many liters does he have?

Solution: $milli = \frac{1}{1{,}000}\ therefore\ \frac{300mL\ x\ 1\not{L}}{1{,}000\not{L}} = 0.3 Liters$

F. Convert ft. to inches: $6\not{ft}\ x \frac{12\ inches}{1\not{ft}} = 72\ inches$

G. Convert 3 lbs. to grams and Kg: $3\not{lbs}\ x \frac{454g}{1\not{lb}} = 1362 grams$

$$kg = 1362\not{g}\ x \frac{1Kg}{1000\not{g}} = 1.36kg$$

H. Convert 3 ounces to grams

Solution: convert ounces to pounds and pound to grams

$$3\not{ozs.}\ x \frac{1\not{lb}}{16\not{oz.}}\ x \frac{454 grams}{1\not{lb}} = 85 grams$$

I. Convert 8,000 ft. to miles and to meters: also kilometers

Solution: convert feet to mile and to meters: also kilometers

$$8000\not{ft}\ x \frac{1\not{mile}}{5280\not{ft}}\ x \frac{1609 meters}{1\not{mile}} = 2437.8 meters = 2.437Km$$

J. Convert 5,000 meters to miles and feet.

Note: 1609 meters is equal to 1 mile and 1 mile is equal to 5280ft.
Solution: meters to miles and miles to feet

$$5000\not{meters}\ x \frac{1 mile}{1609\not{meters}} = 3.10 miles$$

$$feet - 3.10\not{miles}\ x \frac{5280ft}{1\not{mile}} = 16407.7ft.$$

K. Using The Log Table

Finding Logs

What is the log of 4.0?
Find 4 under N or Number in log table. Move horizontally to 0.0 log 4.0 = 0.6

Using log table for root powers

What is the cube root of 81?

Solution: 1. Find log 81, which is 1.908
2. Since we want the cube root, divide by
3. Find inverse log of 0.6361 = 4.3267
Cube root of 81 is **4.3267**

TEMPERATURE SCALES

(Boiling Points of Water)

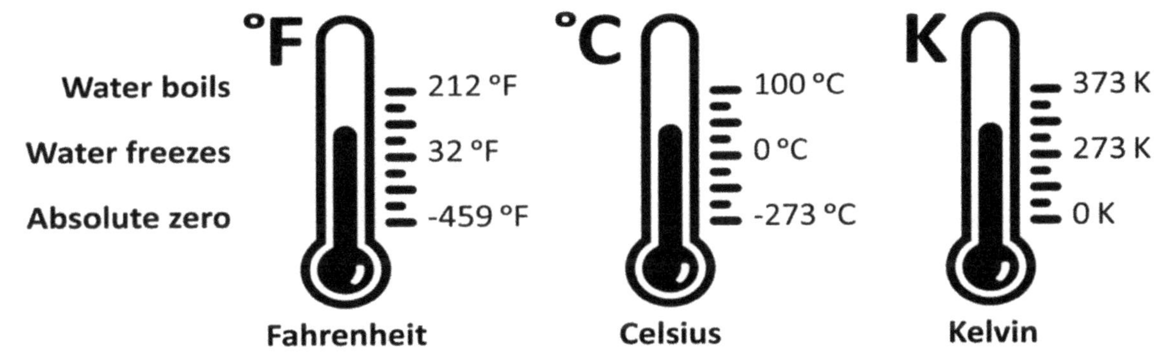

Centigrade to Fahrenheit:
A patient has a temperature of 37°C. What is this in Fahrenheit?

Solution: $°C \ to \ °F = °C \ x \ 1.8 + 32$
$37°C \ x \ 1.8 + 32 = 98.6°F$

Fahrenheit to Centigrade:
An Arizonian once said it was 122°F in the shade. What is it in °C?

Solution: $°F \ to \ °C = F - 32 \div 1.8$
$122 - 32 = 90 \div 1.8 = 50°C$

Absolute or Kelvin:
To convert to absolute simply add °C to 273

Example: $80°C \ to \ °A \ or \ °K$
Solution: $80°C + 273 = 353°K \ or \ °A$

Most Chicagoans will remember December 24, 1983 because it was or below zero. This was a cold night without question. What is in °C?

Solution: A negative number is easy to convert because we do the opposite with 32. An example will simplify this.

Review °F to °C is $F - 32 \div 1.8$ when positive.

However, when working with negative temperatures, we must do the opposite with 32.

Example: Convert -80° to °C (We must find a way to add 32)

Solution: $-80 + -32 = -112$ $\quad$ $-112 \div 1.8 = -62.2$°C

Proof: We could check our calculations by converting -62.18 °C to °F.

Review: $°C\ to\ °F = °C\ x1.8+32$ but using a negative number indicates we must find a way to subtract $32\ °C$.

Therefore, $-62.2°C\ to\ °F = -62.2x1.8+32 = -80°F\ or -62.2x\ 1.8 = -112+32 = -80°F$

LABORATORY MEASUREMENTS

A beginning chemistry student must use the Metric System in the lab to solve chemical calculations. Let's take a look at some common metric terms and calculations.

Volume. Volume refers to liquids and gases. Soda pop, alcohol, oxygen tents, helium tanks, water and swimming pools are common volume references.

Calculations

Volume = Mass/density or M/D

Volume is expressed in liters, milliliters, cubic centimeters (cm^3) and kilometers.

Figure 1-2

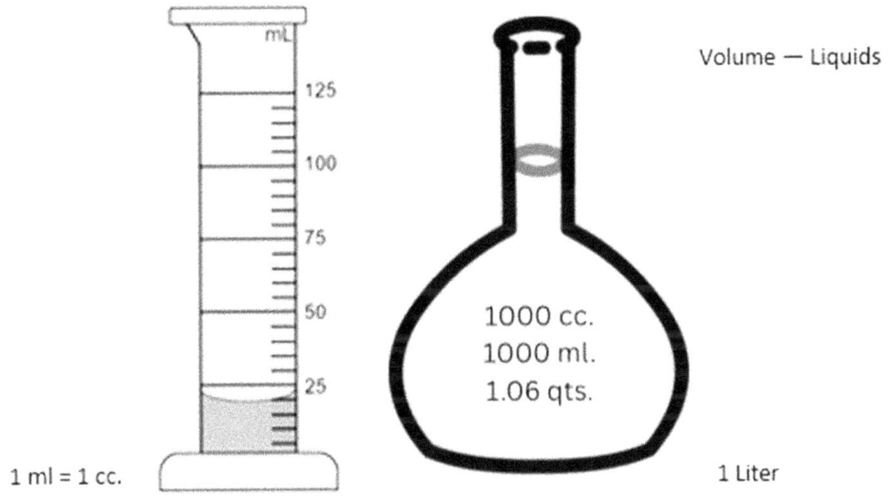

mL
125
100
75
50
25
1 ml = 1 cc.
Volume — Liquids
1000 cc.
1000 ml.
1.06 qts.
1 Liter

The graduated cylinder is used to measure volume

Density. Density is the mass per unit volume: $D = m/v$, example $\frac{grams}{liters}$ or $\frac{grams}{milliliters}$

Mass. Mass is the amount of matter present. M=volume x density or $V\ x$ D.

Example. $Milliliter\ x\ grams\ per\ milliliters{=}mLxg/ml$

Mass Problem

What is the mass of X with 65cc and $6g/cc{=}65cc\ x\ 6g/cc{=}390grams$

Volume Problem

$V{=}M/D$

An object has a mass of 250 grams and a density of $20g/cc$. Find volume.

Solution: $\frac{250\ grams}{20\ g/cc} = 12.5mL$ or $12.5cc$

Density Problem

D=M/V

An object weights 3 grams and has a volume of 12 mL. What is the objects density?

Solution: $\frac{3grams}{12mL} = 0.25g/mL$

General Metric Measurements

Milli = 1/1000	Deci = 1/10	946 mL = 1 quart
Kilo = 1000	454 grams = 1lb (pound)	2.54 cm = 1 inch
Centi = 1/100	1000 meters = 1 Kilometer	1 mL = 1cc or 1cm^3
Deca = 10	1 Kilograms = 2.2 lbs	Mega = 1,000,000

Comparison of SI and English Units of Length

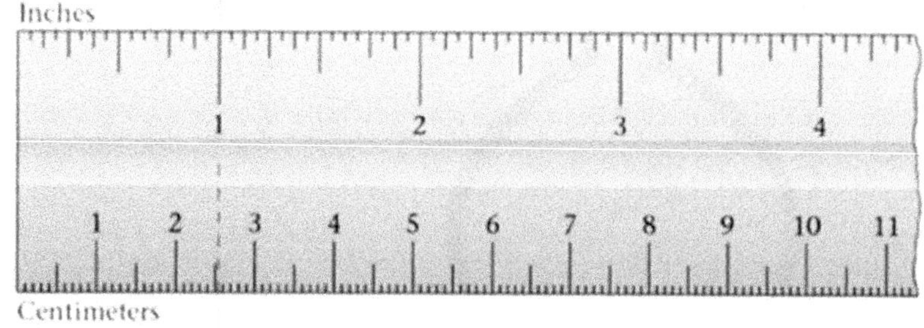

2.54 centimeters (cm) = 1 inch
91.44 centimeters (cm) = 1 yard
1609 meters (m) = 1 mile

Some Common Densities

Symbol		g/cm^3
Au	Gold	19.5
Hg	Mercury	13.5
Pb	Lead	11.3
Fe	Iron	7.86
Al	Aluminum	2.70
H_2O	Water	0.997 or 1.0
	Gasoline	0.7
	Wood	0.5

Volume — solids

1 cm.

1 cm.

1 cm.

(1 cm^3 or 1 cc.)

Note: States of Matter

Solids — definite shape, a definite weight and a definite volume examples: bricks, steel

Liquids — no definite shape but they have a definite weight and volume examples: soda, water

Gases — no definite shape, also molecules move freely, no definite volume but have a definite weight example: air, helium, steam

EXERCISE 1.1

1. 40°c ________ °F ________ °A
2. 4865°F ________ °C ________ °A
3. 38mg ________ grams ________ kg
4. -180°F ________ °C ________ °A
5. 85mg ________ grams ________ kg
6. 125cm ________ mm ________ km
7. An object weighs 4.5 grams and displaces volume of 17 mL. Find density.
8. Sample *X* has a volume of 28cc and a density of 1.75g/cm^3. What is the mass of this object?
9. Object *G* has a mass of 13 grams and a density of 0.45g/cm^3. Find volume.
10. A structure has the following dimensions: length 12, width 4, and height 18. (All in cm_, the density is 12.5g/cc.
11. Solve $2.5x10^{-2}x1.3x10^{-3}$
12. Object T has a volume of 80cm^3 and a mass of 13.6 mg. what is the density of object T?
13. A new candy delight melts at 315°K (or °A). What is this in °F and °C?
14. Convert 16lbs. to Kg.
15. 38.6 inches = 1 meter. How many inches are in 6 meters?
16. Karen won the 10,000 meter race. How many feet, miles and inches did she run?
17. Five ounces equal ________ grams. Note 16oz. = 1lb. and 45grams = 1lb.
18. Patient X weighs 54,480grams. How many pound does she weigh?
19. Dr. Ray give the infant Sam took three (3) 25mg tablets of iron for 2 days. How many grams did the infant Sam get? (Note he only took 3 per day.)
20. Solve the following:

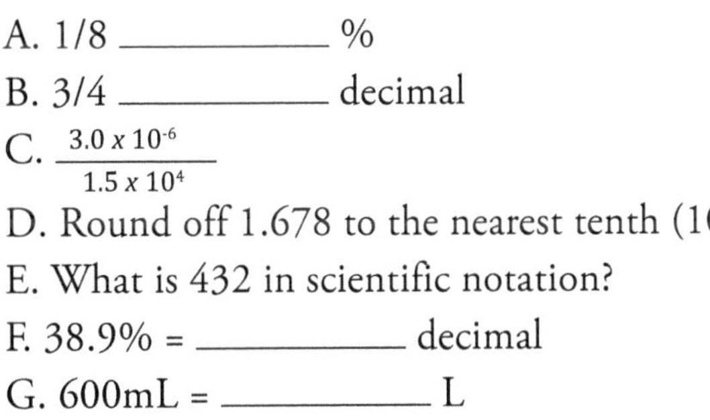

A. 1/8 ________ %

B. 3/4 ________ decimal

C. $\frac{3.0 \times 10^{-6}}{1.5 \times 10^{4}}$

D. Round off 1.678 to the nearest tenth (10th)

E. What is 432 in scientific notation?

F. 38.9% = ________ decimal

G. 600mL = ________ L

H. 3.25mL = ________ mL

21. The density of gas X a $25°C\ is\ 1.6x10^{-4}g/cm^3$. What is the volume of 35g of gas X?
Note: V = M/D

22. A tank is full of an experimental fluid. A chemist's records a mass of 1,350 grams and a density of $1.78g/cm^3$. Find the volume.

CHAPTER 2

Molar Mass

Objectives and Goals

To be able to:

- use the periodic table to calculate molar mass of compounds
- learn a few symbols

Keywords

- molas mass
- atomic mass unit
- gram molecular weight
- formula weight
- weight
- mass

CHAPTER 2

MOLAR MASS

A chemist must know how to calculate the molar mass of a compound. This is easily done by adding the weights of the elements in the compound. Note: Formula Mass can be written in **a.m.u.** = atomic mass units.

Example 1

What is the formula mass or mola mass of NaCl (sodium chloride) or table salt?

Solution: Mass of Na = 23 grams x 1 = 23
Mass of Cl = 35.5 grams x 1 = 35.35

58.5 grams or 58.5 a.m.u.

Adding both masses, we get 58.5 grams or one (1) formula mass of NaCl.

Example 2

What is the molas mass of $Al_2(SO_4)_3$ aluminum sulfate?

Solution: **Mass x No.**

Al = 27 x 2 = 54 grams
S = 32 x 3 = 96 grams
O = 16 x 12 =192 grams

$Al_2(SO_4)_3$ mass = 342 grams or 342 a.m.u.

Example 3

What is the molas mass of *baking soda* $NaHCO_3$ (sodium bicarbonate)?

Solution: **Mass x No.**

Na = 23 x 1 = 23 grams
H = 1 x 1 =1 gram
C = 12 x 1 = 12grams
O = 16 x 3 = 48 grams

$NaHCO_3$ mass = 84grams or 84 a.m.u.

EXERCISE 2-1

Calculate the formula mass of the following: (Refer to the periodic table for element weight)

- $MgCO_3$
- $C_6H_{12}O_6$
- $KClO_3$
- H_2O
- $AgNO_3$
- KCl
- $C_{12}H_{22}O_{11}$
- $Ba(NO_3)_2$
- $Ni(NO_3)_2$
- $Bi(OH)_3$
- $Ca(ClO_3)_2$
- H_2SO_4
- $Na_2Cr_2O_7$

EXERCISE 2-2

Calculate the formula mass of the following: Use your own paper.

- Fe_2O_3 (Rust)
- $C_{12}H_{22}O_{11}$
- KCN
- H_2CO_3
- $MgSO_4$
- $AsCl_5$
- $PbSO_4$
- AgCl
- KOH
- $C_6H_8O_6$ (Vitamin C)
- N_2H_4CO (Urea)
- CH_3OCH_3 (dimethyl ether)

CHAPTER 3

The Mole Concept

Objectives and Goals

To be able to:

- solve for molar mass
- convert grams to moles
- convert moles to grams
- study the Avogadro number

Key words

- Avogadro's number
- mole
- molar mass

CHAPTER 3

THE MOLE CONCEPT

The gram formula mass of a compound is conveniently expressed as mole.

Example 1:

How many grams in 0.5 moles of **$KClO_3$**?

Solution: First we calculate the mass of $KClO_3$

K = 39 x 1 = 39

Cl = 35 x 1 = 35.5

O = 16 x 3 = 48

---

1 mole = 122.5grams or 122.5 a.m.u.'s

$$0.5\cancel{moles}\ KClO_3\ x\ \frac{122.5g}{1\ \cancel{mole}\ KClO_3} = 61.25grams$$

Example 2:

How many grams in 0.187 moles of $C_6H_{12}O_6$ (glucose) ?

Solution: C = 6 x 12 = 72

H = 12 x 1 = 12

O = 6 x 16 = 96

180grams or 1 mole of $C_6H_{12}O_6$

Therefore: $0.187\cancel{moles}\ x\ \frac{18grams}{1\cancel{mole}\ C_6H_{12}O_6} = 33.6grams$

Example 3:

How many moles in 60 grams of $CaCO_3$?

Solution: Ca = 40 x 1 = 40

C = 12 x 1 = 12

O = 16 x 3 = 48

---

100 grams

Since the formula mass of $CaCO_3$ is 100, grams and/or one mole; 100 is the denominator; therefore , 60 grams to mole is:

$$60\ \cancel{grams}\ x\ \frac{1mole\ CaCO_3}{100\ \cancel{g}} = 0.6\ moles\ of\ CaCO_3$$

Example 4:

A chemist has 12.8 grams of $AgNO_3$. How many moles does he have?

Solution: The formula mass of $AgNO_3$ is: Ag = 108 x 1 = 108 a.m.u.

N = 14 x 1 = 14 a.m.u.

O = 16 x 3 = 48 a.m.u.

1 mole = 170 a.m.u.

$$\frac{12.8g/1mole}{170\ AgNO_3} = \frac{12.8\cancel{g}}{170\cancel{g}} = 0.0753\ moles\ of\ AgNO_3$$

AVOGRADO'S NUMBER

6.023 x 10^{23} particles = 1 mole of a substance

One mole of a compound equals 6.023 x 10^{23} particles. How many particles in 0.5 moles in $CaCO_3$?

Solution: $0.5\ x\ \frac{6.023\ x\ 10^{23}}{1\ mole\ particles} = 3.011\ x\ 10^{23}\ particels$

Convert to A-F to moles.

A. 12 grams of $KClO_3$

B. 7.8 grams of H_3BO_3

C. 380 grams of $C_6H_{12}O_6$

D. 0.93 grams of $NaHCO_3$

E. 1.65 kg of K_2CO_3

F. What is Avogadro's Number of A, B, and C? *Convert G-I to grams:*

G. 0.087 mole of MgO

H. 2.85 moles of HgO

I. 134 moles of NaOH

J. **FOR FURTHER STUDY:**

How many grams in 1 particles of S?

Solution: S = 32 grams or 1 mole = 32 grams
(note: we must start with Avogadro's Number)

Avogadro's Number = $(N_A) = \frac{1.0 \, x \, 10^0 \, x \, 1 \, mole}{6.0 \, x \, 10^{23}} = 1.66 \, x \, 10^{-24}$ moles of sulfur
Moles of sulfur to grams of sulfur is;

$$moles \; x \; wt. of \; sulfur = 1.66 \, x \, 10^{-24} \, x \, 1 \, mole \frac{32g \; sulfur}{1 \, mole \; sulfur}$$
$$= \mathbf{5.3 \, x \, 10^{-23}} grams \; of \; sulfur \; in \; one \; paticle$$

Proof: Convert our answer in grams to one particle:

$$\cancel{5.3x10^{-23}g} \, x \, \frac{1 \, mole}{\cancel{32g}} \, x \, \frac{6.02x10^{23}}{1 \, mole} = 1 \, particle$$

K. How many atoms in 17 grams of Na?

Solution: $17 grams \; of \; Na \, x \, \frac{1 \, mole \; Na}{23g} x \frac{6.02x10^{23}}{1 \, mole \; atoms \; Na} = 4.45 \, x \, 10^{23} \; atoms$

FOR FURTHER STUDY:

Solve the following:

L. 6 grams of Ca to moles __________ to atoms __________
M. 3.0×10^8 atoms of Ag to grams of Ag __________
N. 1.5×10^{23} molecules of H_2S __________ moles
O. 3.5×10^{-2} moles of Cl_2 __________ moles
P. 17.5 grams of $AgNO_3$ __________ moles __________ atoms

CHAPTER 4

Percent Composition

Objectives and Goals

To be able to:

- find the percent composition of a compound
- apply percent composition concepts to moles, mass and ratios

Key words

- ratios
- percent
- mass
- formulas
- compounds
- moles

CHAPTER 4

PERCENT COMPOSITION

Percent composition is self-explanatory. A chemist can calculate percentage by using the formula part over the whole. A few examples will simplify this.

EXAMPLE 1:

What is the percent composition of H_2O

First we calculate the formula mass of H_2O: H = 2 x 1 = 2

O = 1 x 16 = 16

---

18 grams = 1 mole of water (H_2O)

$$\% \text{ of H} = 2\cancel{g}\ x\ \frac{1 mole}{18\cancel{g}\ H_2O}\ x\ 100 = 11.11\%$$

$$\% \text{ of } 0 = 16\cancel{g}\ x\ \frac{1 mole}{18\cancel{g}\ H_2O}\ x\ 100 = 88.8\%$$

EXAMPLE 2:

What is the percentage of iron in rust?

The formula is Fe_2O_3

The mass is calculated as follows:

Fe = 2 x 56 = 112

O = 3 x 16 = 48

Formula mass = 160grams % of Fe 112/160 = 0.7 or 70%

$$Fe = 112\cancel{g}\ x\ \frac{1\ mole}{160\cancel{g}\ Fe_2O_3}\ x\ 100 = 70\%$$

EXAMPLE 3:

How many grams of Al in 25 grams of Al_2O_3?

Solution: Mass of Al_2O_3 is 102 grams (Al = 2 x 27 = 54 and 0 = 3 x 16 = 48)

$$\frac{25g\ Al\ x\ 27g\ x\ 2\cancel{moles}\ x\ \cancel{1mole}\ Al_2O_3}{\cancel{1\ moles\ Al}\ x\ 102\ g} = 13.23g\ of\ Al$$

Example 4:

How many grams of Ag in 0.35 moles in $AgNO_3$?

Step 1 find % of Ag in the compound

$$\%\ of\ Ag = \frac{wt.\ of\ Ag\ 108\ grams\ x\ 100}{1\ mole\ of\ AgNO_3\ 170grams} = 63.5\%\ of\ Ag$$

Step 2: grams of Ag = $\frac{\%\ of\ Ag\ 0.635\ x\ 0.35\ \cancel{moles}\ x\ 170\ grams}{1\ \cancel{mole}\ AgNO_3} = 37.78\ grams\ of\ Ag$

Figure 4-1

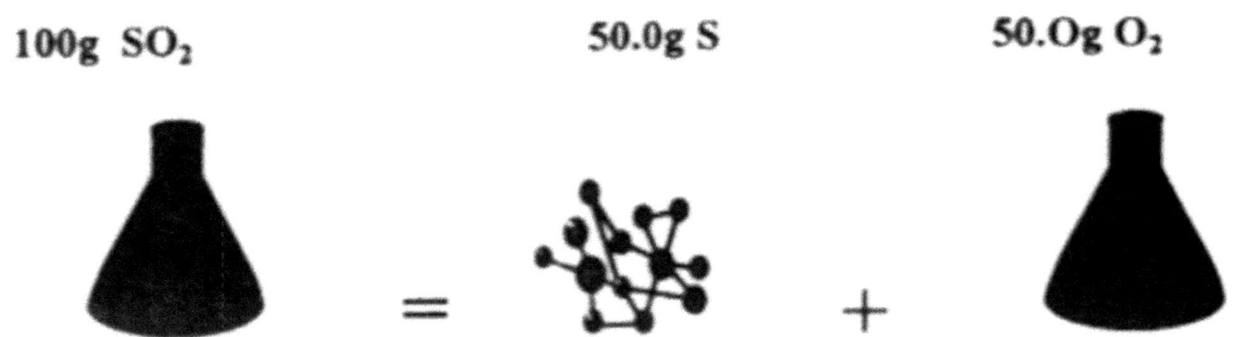

Percent analysis of SO_2 (wt = 64g) $S = \frac{32}{64} x\ 100 = 50\%$ weight of compound

$O = \frac{32}{64} x\ 100 = 50\%$ weight of compound

Therefore, as is shown in Figure 4-1, 100g of SO_2 = 50g S and 50g O_2. (Note later we will connect 50g of O_2 to liters).

Exercise 4-1

(Percentages)

1. What is the percentage of oxygen in soda pop or carbonic acid?

   H_2CO_3

2. What is the percentage of Na in soda ash?

   Na_2O_3

3. What is percentage of the antacid magnesium hydroxide (Milk of Magnesia)?

$$Mg(OH)_2$$

4. Calculate the percentage of the underlined.

$\underline{Ca}CO_3$ Pb$\underline{(NO_3)_2}$ $\underline{Fe_2}O_3$

5. Calculate the percentage of the underlined.

% of $\underline{O_2}$ in 0.45 moles of H_2O_2

6. Grams of ***Cu*** in 0.16 moles in **$CuSO_4$**
7. What is the % of ***Al*** in **$Al_2(SO_4)_3$**
8. A chemist has 60 grams of **$Al_2(SO_4)_3$**. How many grams of ***Al*** does he have?
9. What is the percent composition of **$C_{21}H_{30}O_2$**. (The active ingredient in marijuana?)
10. Assuming it is possible to extract carbon from **$C_{21}H_{30}O_2$**, how many grams of ***C*** can be extracted from 62.8 grams of the marijuana compound?
11. Which compound has the most ***Na*** present?

$NaHSo_4$ **$Na_2S_2O_3$** **$Na_2S_2O_4$**

12. A designer ring weighs 60 grams. The formula is **$AuCl_2$**. The ring has ________ grams of gold.
13. What is the % of ***Pt*** (platinum) in the anti-cancer drug Platinol **$Pt(NH_3)Cl_2$**?
14. Teflon is produces by the polymerization of tetrafluoroethylene, **$CF_2 = CF_2$**: What is the % of ***C*** and ***F*** in Teflon?
15. Formaldehyde **HCHO** is used to preserve biological specimens and embalming fluid. A scientist has **80 ml** of **HCHO**. How many mL of ***C*** and ***H***?
16. Find the percentage composition of nitroglycerin. **$C_3H_5N_3O_9$**
17. Find the % composition of penicillin **$C_{14}H_{20}N_2S0_4$**
18. How many grams of C in 40 grams of penicillin?
19. Honeybees produce a pheromone (an alarm or warning chemical) *isoamyl acetate* **$C_7H_{14}0_2$**. How many grams of ***oxygen*** in 19 grams of *isoamyl acetate*?
20. Toys, plastic piping, plastic bottles and floor tile are made using PVC or polyvinylchloride. Calculate the % composition **of PVC (CH_2CH) n**

Cl

21. Caffeine is a stimulant found in coffee, tea and several sodas. A college student has 30 grams of caffeine. How many grams of **C.N.O.** and **H** does she have?

Figure 4-2

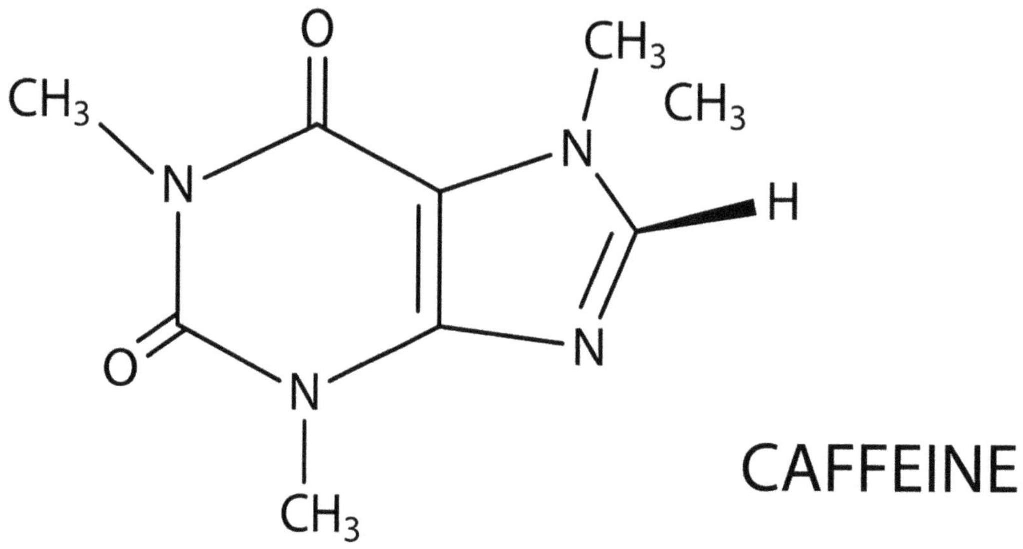

CH3
O
CH3
CH3
N
N
H
O
N
N
CH3

CAFFEINE

CHAPTER 5

Empirical Formulas

Objectives and Goals

To be able to:

- solve for empirical formulas
- use percent as analytical tool
- name empirical formulas
- study mole ratios

Keywords:

- mole
- percent composition
- empirical

CHAPTER 5

EMPIRICAL FORMULAS

Empirical Formula (Simplest formula)

The empirical formula is often called the simplest formula of a compound. In other words, the compounds can only be divided or multiplied by one. Furthermore in an empirical formula we get the smallest whole number ratio among elements.

Glucose empirical is CH_2O Ratio 1:2:1

EXAMPLE 1

An unknown compound contains 25% H and 75% C. What is the empirical formula?

Solution: %

Step 1: wt. of element

Convert % to decimal

$$H = \frac{0.25 \times 1 \text{ mole } H}{1 \cancel{g}} = 0.25 \; mol \; of \; H \; atoms$$

$$C = 0.75 \times \frac{1 \text{ mole } C}{12 \cancel{g}} = 0.0625 \; mol \; of \; C \; atoms$$

Step 2: Divide moles by least common denominator or smallest atom mol (mole) ratio which is 0.0625 mol of C atoms.

$$C = \frac{0.0625}{0.0625} = 1 \; mol \; ratio \; of \; C \; atom$$

$$H = \frac{0.25}{0.0625} = 4 \; mol \; ratio \; of \; atom$$

1 C atoms and 4 H atoms = CH_4 is the empirical formula.

EXAMPLE 2

A chemist has 53% C and 47% O. Find the empirical formula.

Solution:

$$C = 0.53 \times \frac{1 \; mole \; Carbon}{12 \cancel{g}} = 0.04416 \; mole \; of \; C \; atoms \; mole \; ratio$$

$$O = 0.47 \times \frac{1 \; mole \; Oxygen}{16 \cancel{g}} = 0.0293 \; mole \; of \; O \; atoms$$

Divide both ratios by the smallest number which is 0.0293

$$C = \frac{0.04416}{0.0293} = 1.5\ atom$$

$$O = \frac{0.0293}{0.0293} = 1\ atom$$

Since atoms must be whole number and C is 1.5 we double both numbers

C = 1.5 x 2 = 3 atoms

O = 1 x 2 = 2 atoms

Formula C_3O_2

EXAMPLE 3:

A chemist burns 3.445 grams of Al in air. The final mass of the oxide is 6.5 grams. What is the empirical formula of the compound?

Solution: Grams over weight of elements and divide by the smallest number (atom mole ratios).

Al = 3.445 grams 0 = 6.5 grams – 3.445 grams = 3.055 grams

Al = 3.445 grams 0 = 6.5 grams - 3.445 grams = 3.055 grams

$$Al = 3.445 \cancel{grams}\ x\ \frac{1\ mole\ Al}{27\cancel{g}} = 0.1276\ mole\ ratio\ Al\ atoms$$

$$0 = 3.055 \cancel{grams}\ x\ \frac{1\ mole}{15.999\ \cancel{g}} = 0.1909\ mole\ ratio\ O\ atoms$$

Dividing both mole values by the smallest atom mole ratio equals: (0.1276)

$$Al = \frac{0.1276}{0.1276} = 1\ atom\ mole\ ratio \qquad 0 = \frac{0.1909}{0.1276} = 1.50\ atom\ mole\ ratio$$

Since compound must contain whole numbers of atoms, our empirical formula should contain only whole numbers. Therefore, since Oxygen is 1.5, both elements must be multiplied by 2.

Empirical Formula: Al = 1 x 2 = Al_20_3

0 = 1.5 x 2

EXERCISE 5-1

A. 20% Ca, 80% Br ____________

B. 56.4% Mn, 43.6% S ____________

C. 23.1% Al, 15.4% C, 61.5% O ____________

D. 39.6% Na, 0.6034% Cl ____________

E. 11.1% H, 88.8% O ____________

F. A compound consists of 10.05% C, 0.84% H, and 89.11% Cl. What is its empirical formula?

G. A chemist burns 4.857 grams of Ca (calcium) and has a final weight of 6.8 grams. What is the empirical (simplest) formula for this oxide?

H. A chemist has 1.5 grams of a compound. Analysis shows 0.5 moles of Mg and 1.00 moles of N. What is the empirical formula?

I. When 8.75g of platinum chloride is heated, chlorine escapes and 6.06g of platinum is left. What is the empirical formula?

J. A common commercial gas is used to make cleaning solutions. This gas is 82.35% N and 17.64% H. what is the empirical formula.

CHAPTER 6

Molecular Formulas

Objective and Goals

To be able to:

- solve for molecular formulas
- compare empirical and molecular formulas

Key terms

- molecular formula
- mole
- ratios
- significant figures
- empirical formulas

CHAPTER 6

MOLECULAR FORMULAS

The following are needed to calculate the molecular formula of a compound:

- % composition
- weight of elements
- molecular weight of compound
- indicates atoms present

The molecular formula differs from the empirical formula because we must have the molecular weight to calculate a molecular formula.

$$\textbf{Molecular Formula} = \frac{\textbf{Molecular Formula weight}}{\textbf{Empirical Formula weight}}$$

Example 1

An organic compound consists of 92.95% carbon and 7.75% hydrogen. The molecular mass is 78 grams. What is the molecular formula? Note: gram atomics weights of C = 12 and H = 1

Solution: *note convert percent to grams*

Step 1:

$$C = \frac{92.95\cancel{g} \; x \; 1 \; mole \; C}{12\cancel{g}} = 7.7 \; moles \; C$$

$$H = \frac{7.75\cancel{g} \; x \; 1 \; mole \; H}{1\cancel{g}} = 7.75 \; moles \; H$$

Step 2: Divide by the smallest moles since 7.7 moles is the smallest we have:

$$C = \frac{7.7}{7.7} = 1 \; mole \quad H = \frac{7.75}{7.7} = 1.006 = 1 \; mole$$

Since molecular formula

$$\frac{Molecular \; Weight = 78grams}{Empirical \; Weight = 13 \; grams} = 6CH$$

Molecular Formula = $\mathbf{C_6H_6}$ *Benzene*

EXERCISE 6-1

Solve

A. 43.6% P, 56.4% O, M.W = 284

B. 3.65% H, 37.8% P, 58.5% O, M.W 164 = grams

C. 84.2% C, 15.78% H, M.W = 228 grams

D. 46.6% Si, 53.3% O, M.W = 180 grams

E. 25.9% N, 74.07% O, M.W = 432 grams

F. An astronaut collects an unknown compound. Analysis shows that two elements are present. Element X is 20% and weighs 17 grams. Element Y is 80% and weighs 68 grams. M.W is 85 grams. Find the molecular formula.

G. Define Key Terms

- Empirical formula
- Molecular formula
- Molar mass
- Mole
- Significant digits

CHAPTER 7

Chemical Equations

Objectives and Goals

To be able to:

- balance chemical equations
- to study mole ratios
- to apply basic math concepts

Key words

- coefficients
- reactants
- products
- yields
- direct combination reactions
- displacement reactions
- decomposition reactions

CHAPTER 7

CHEMICAL EQUATIONS

Chemical reactions are often defines as a change in the compositions of matter. Matter is anything that contains weight and occupies space and is involved in chemical reactions.

Example of Desired Reactions

1. Conversion of O_2 to CO_2 in animals (respiration)
2. Production of soda is $H_2O + CO_2 \rightarrow H_2CO_3$
3. Production of O_2 for oxygen tents
4. Food preservatives, cooking, digestion
5. Artificial flavors; powdered soft drinks; plastics, synthetic clothing
6. Solubilities of medicines in the body
7. VHS tapes, batteries, radios, television, etc.

Example of Undesired Reactions

1. Rusty bicycle or rust $\mathbf{Fe_2O_3}$(s)
2. Nuclear fallout and radiation
3. Air pollution
4. Carbon Monoxide destroys animal respiration
5. Narcotics

Moreover, any scientist could have an endless list of chemical reactions. Since there are countless reactions, chemists use equations and classify them based on the changed in the equations.

Remember, in chemical reactions molecules either move together or move apart and usually involve energy.

BALANCING CHEMICAL EQUATIONS

Chemical Equation Symbols	
(s)	= solid
(g)	= gas
(l)	= liquid
(aq)	= water or liquid
↑	= given off
↔	= reversible

Note: Both sides must be equal. Study the subscripts and co-efficients for accuracy.

Example 1 $H_{2(g)} + N_{2(g)} \quad 2NH_{3(g)}$
reactants ***products***

The above equation is unbalanced because hydrogen has a subscript of 3 on the product side and a co-efficient of one on the reactant side. Therefore, we need two moles of NH_3.

Balanced

$$3H_{(2)(g)} + N_{(2)(g)} \rightarrow 2NH_{3(g)}$$

Example 2 $2CaCO_{3(2)} \rightarrow CaO_2 + CO_{2(g)}$

Hint The co-efficient 2 for $CaCO_3$ and subscript O_2 in CO_2 counting the elements in $CaCO_3$ we have 2Ca, 2 C's and O_6 or 2 x O_3.

Balanced

$$2CaCO_{3(s)} \rightarrow 2CaO_{(s)} + 2CO_{2(g)}$$

TYPES OF EQUATIONS

A. Direct Combination

Two elements combine to form a compound (e.g., formation of water)

$$2H_{(2)(g)} + O_{(2)(g)} \rightarrow 2H_2O_{(l)}$$

B. Displacement

One or more of the elements change places like the compounds as Zn does in the production of hydrogen gas.

$$Zn_{(s)} + 2HCl_{(l)} \rightarrow ZnCl_{2(l)} + H_{2\updownarrow(g)}$$

C. Decomposition

A compound is usually heated to form two or more compounds as in the decomposition of potassium chlorate to form potassium chloride and oxygen.

$$2KClO_{3(s)} \rightarrow 2KCl_{(s)} + 3O_{2\uparrow(g)}$$

D. Redox

The exchange of positive and negative charges or ions will be discussed in detail later.

EXERCISE 7-1

(Balance the following equations)

Direct Combination

A. $Ca_{(s)} + O_{2(g)} \rightarrow CaO_{(s)}$
B. $P_{(s)} + Cl_{2(g)} \rightarrow P_cl_{3(g)}$
C. $Hg_{(s)} + O_{2(g)} \rightarrow HgO_{(s)}$
D. $Mg_{(s)} + N_{2(g)} \rightarrow Mg_3N_{2(s)}$
E. $P_{(s)} + 50_{2(g)} \rightarrow P_2O_{5(s)}$

Displacement

a. $Br_{2(g)} + Kl_{(s)} \rightarrow KBr_{(s)} + I_{2(g)}$
b. $Cl_{2(g)} + NaBr_{(s)} \rightarrow NaCl_{(s)} + Br_{2(g)}$
c. $Zn_{(s)} + HCl_{(1)} \rightarrow ZnCl_{2(1)} + H_{2(g)}$
d. $Zn_{(s)} + AgNO_{3(s)} \rightarrow Zn(NO_3)_{2(s)} + Ag_{(s)}$
e. $Mg_{(s)} + H_2SO_{(l)} \rightarrow MgSO_{4(s)} + H_2\boxed{\nearrow}$

DISPLACEMENT

$Zn + 2HCl \rightarrow ZnCl_2 + H_2\boxed{\nearrow}$

Note formation of H_2 gas.

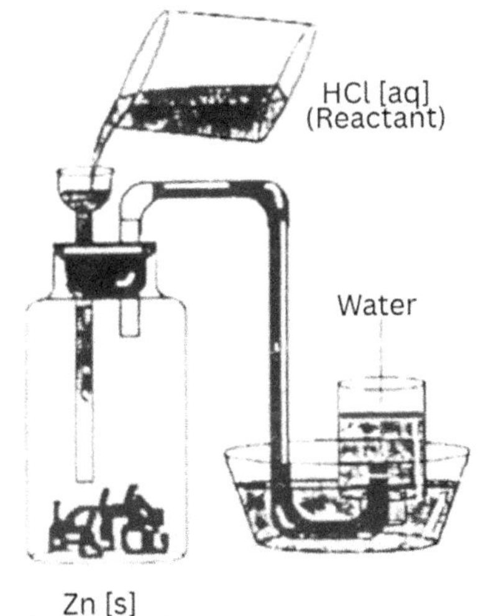

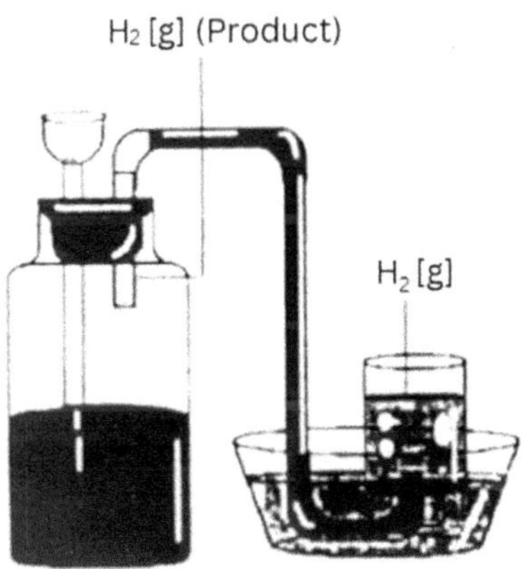

ZnCl [aq]
(Product)

Decomposition

a. $NaNO_{3(s)} \xrightarrow{heat} NaNO_{2(s)} + O_{2(g)}$

b. $HNO \xrightarrow{heat} HNO_3 + NO_{(g)} + H_2O_{(l)}$

c. $CCl_{4(l)} \xrightarrow{heat} C_{(g)} + CL_{2(g)}$

d. $KClO_{3(s)} \xrightarrow{heat} KCl_{(s)} + O_{2(g)}$

e. $2AlCl_{3(s)} \xrightarrow{heat} Al_{(s)} + Cl_{2(g)}$

Diagram A

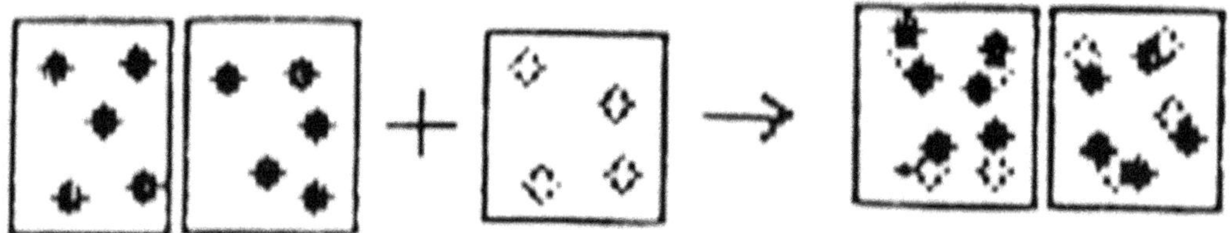

Note the formation of water. Each block is one mole, so 2H2 + O_2 2 moles of H_2O.

FOR FURTHER STUDY

1. Describe the Diagram below in moles (reactants and products).

Diagram B

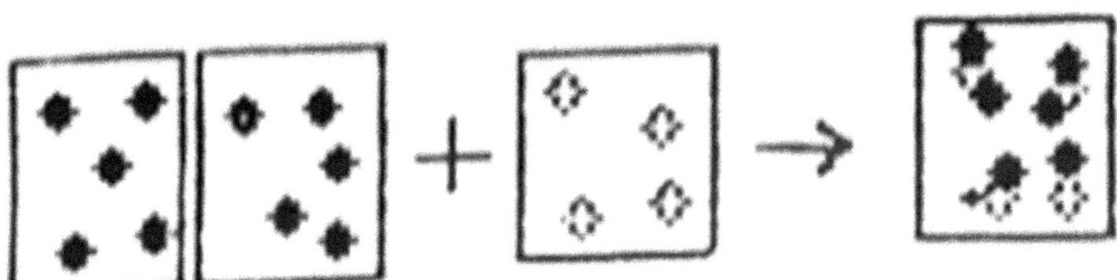

2. Write an equation that would male oxygen. What type of equation (displacement, decomposition, or substitution)?
3. Balance the formation of carbonic acid (soda).

$$H_2O + CO_2 \rightarrow$$

4. Photosynthesis (balance the equation)

$$6CO_2 + H_2O \text{ chlorophyll } \xrightarrow{light} C_6H_{12}O_6 + O_2 + 6H_2O$$

List the reactants below: ____________________

____

____

List the products below: ____________________

____

____

5. Chemist have used chemical reactions for countless purposes. In this case a technician passes an electric current through H2O.

$$2H_2O \text{ electric current } H_2 + O_2$$

   a. Balane the equation above.
   b. Give the number of moles for each reactant and product.
   c. How can welders, doctors, divers and rocket engineers use this equation?

Remember: Study the co-efficient when balancing chemical equations.

CHAPTER 8

THE HYDRATE

Goals and Objectives

To be able to:

- review the empirical formula concepts
- solve for hydrates
- apply percentage concepts to hydrate problems

Keywords:

- hydrates
- percent
- crystal lattice
- formulas

CHAPTER 8

THE HYDRATE

Another kind of formula used to represent compounds is called a hydrate. A hydrate is an ionic compound that contains water molecules within crystal lattice. Some examples are **gypsum $CaSO_{4+}$ $2H_2O$** and **$CaCl_2$ $6H_2O$**.

EXAMPLE 1

Hydrate Problem

A chemist heats 15 grams of $BaCl_2$. His final weight is 12.78 grams. Find the hydrate $BaCl_2$ x $2H_2O$

$$BaCl = \frac{12.78\ \cancel{grams}\ x\ 1\ mole\ BaCl_2}{208\ \cancel{grams}} = 0.06144\ moles$$

Solution: H_2O = 15grams - 12.78grams = 2.22 grams

$$H_2O = \frac{2.22\ \cancel{grams}\ x\ 1\ mole\ H_2O}{18\ \cancel{grams}} = 0.1233\ moles$$

Divide both compound by the smallest mole ration which is 0.06144

$$BaCl = \frac{0.06144}{0.6144} = 1\ mole \qquad H_2O = \frac{0.1234}{0.6144} = 2\ moles$$

Hydrate formula is $BaCl_2 \bullet 2H_2O$

EXAMPLE 2

Hydrate

65% $MgSO_4$ 1 mole $MgSO_4$= 120 grams

34.7 H_2O 1 mole H_2O = 18 grams

Solutions: (note we will convert % to decimals)

$$MgS0_4 = \cancel{0.65}\ x\ \frac{1\ mole\ MgS0_4}{\cancel{120}\ g} = 0.0542\ moles$$

$$H_20 = \cancel{0.347}\ x\ \frac{1\ mole\ H_20}{\cancel{18}\ g} = 0.01927\ moles$$

H_20 = 0.347 $MgSO_4$ = 0.01927

Dividing by the smallest ration 0.0542

$$MgSO_4 = \frac{0.0546}{0.0546} = 1\ mole\ ratio \qquad H_20 = \frac{0.01927}{0.0546} = 3.5\ mole\ ratio$$

Note: We multiply by 2 because of 3.5 ratio.

Hydrate is 2$MgSO_4$ x 7 H_2O

Exercise 8-1

Hydrate Problems

1. $CuSO_4$ x H_2O, 39% H20 and 60.86% $CuSO_{4}$.
2. $CoCl_2$ x H_2O 78.3% $CoCl_2$and 21.67% H_2O
3. $MgCO_3$ x H_2O, 60.87% $MgCO_3$ and 39.13% H_2O
4. Na_2CO_3 x H_2O, 20 grams of Na_2CO_3 is heated with a final result of 12.58 grams of H_2O and 7.41 grams of Na_2CO_3

Key terms — define

- hydrate
- crystal lattice
- ionic

CHAPTER 9

The Periodic Table

Goals and Objectives

To be able to:

- study basic periodic laws
- use the periodic table as a helpful tool
- study groups, periods, ionization energy and trends
- study atomic numbers

Keywords:

- alkali metals
- alkaline metals
- atomic numbers
- atomic mass
- atomic radii
- electronegativity
- groups
- halogens
- inert gases
- ionization energy
- metalloids
- periods
- actinide series
- lanthanide series
- periodic law
- transition elements

CHAPTER 9

PERIODIC TABLE TERMS

PERIODIC TABLE TERMS

- Protons – a positive charge and atomic number
- Electrons – negative charge
- Neutrons – neutral
- Electronegativity – attraction for electrons to form bonds
- Atomic radius – one-half the distance between two nuclei or internuclear distance
- Orbitals – a region outside the atom in which electrons can be found
- Ions – transfer of electrons resulting in charge particles
- Coinage elements – god, silver, copper

PERIODIC TABLE

The periodic table is an invaluable tool for scientist, chemists, physicists and technicians. Every lab has one because it is used for countless purposes.

Moreover, the table is like a telephone book in that scientists simply look up symbols and weights as we would telephone numbers. Technicians use the table to predict chemical reactions and most chemists and physicist use the table to study elements and their properties especially in various compounds.

The periodic table is important to chemists because of its vital information.

Elements are functions of their atomic numbers or numbers of protons. The mass is equal to number of protons and neutrons.

GROUPS

Groups 1, 2 and 3 (metals and metalloids) are vertical rows. As we proceed down, group metals become more active and atomic radius increases.

Groups 4, 5, 6 and 7 gases on the other hand become less active and acquire darker color as we proceed down.

Group 8 are noble which are inert and do not react with most elements.

PERIODS

Periods are horizontal. In general, metallic activity decreases as we move horizontally while gas activity increases. (Example: *Na* is more active than *Mg* and *F* is more active than *0.*)

OXIDATION NUMBERS

Metals tend to lose electrons equal to their group number. This result in +1 for group 1, +2 for group 2, and +3 for group 3 and -4 for group 4.

Non-metals tend to gain electrons. Group 5 is -3 because it can gain 3 electrons that total 8, which is the maximum for the outer shell.

Moreover, Group 6 is -2, group 7 is -1 and Group 8 is 0. Note for non-metal, the oxidation numbers can be calculated by subtracting the group from eight.

RADIUS

The atomic radius increases as we go down a group because more electrons shells are added. On the other hand, radius decreases as we go across a period because we are adding electrons to the same shell.

IONIC

For metals, the ionic radius is less than atomic radius. This is due to the positive metal charge which pulls in the shells, causing a decrease in ionic radius. For non-metals, the ionic results is greater than the atomic radius because the weal positive charge does not pull the shells in closer to the nucleus.

TRANSITIONAL ELEMENTS

When we look at the periodic table, we see elements labeled B. these are the transitional elements. They lose electrons from both inner and outer shells. Moreover, transitional elements give off colors and many have 2 electrons in their outermost shell.

STATE OF MATTER

Some periodic tables use color to indicate solids, liquids, and gases. The lighter the color, the more unstable. (Example: black-solid, blue-liquid, and orange-gases.)

PERIODIC LAW

Elements are properties of their atomic numbers and generally increase by atomic numbers.

EXERCISE 9-1

(Periodic Review)

1. Most chemists would agree that Li is more active than Be. Why is this true?
2. How do transitional elements differ from non-transitional elements?
3. Describe the activity of elements in groups and periods?
4. Describe the atomic radius and ionic radius in groups and periods. Explain what causes these differences.

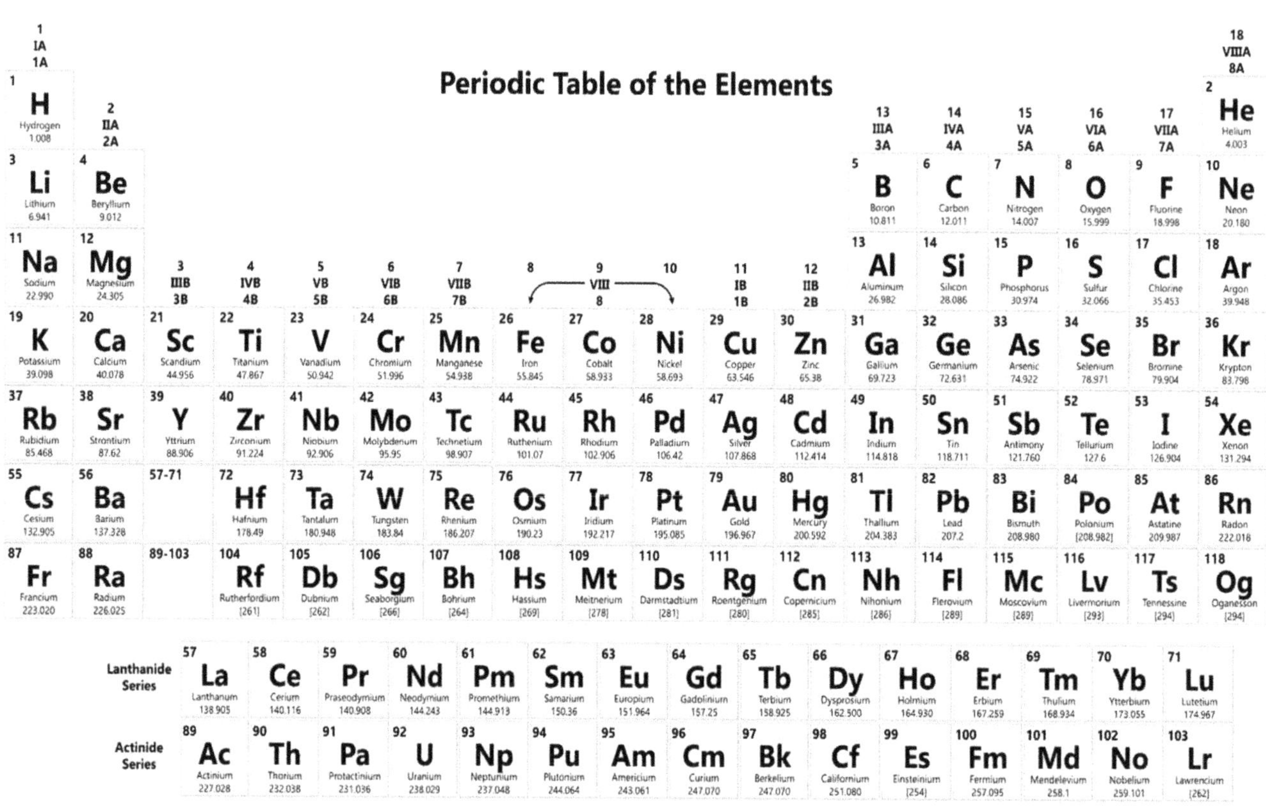

CHAPTER 10

Thermochemistry

Goals and Objectives

To be able to:

- solve for calories
- study heat capacity
- study basic water constants
- convert form calories to joules
- convert joules to calories
- to find an unknown specific heat

Keywords: (refer to any science book or dictionary)

- specific heat
- heat capacity
- freezing point
- boiling point
- heat of vaporization
- calories
- joules
- thermochemistry
- energy

CHAPTER 10

THERMOCHEMISTRY

Thermochemistry is the study of heat reactions. An example of temperature reaction is a steam bath versus a sauna. Some athletes claim the steam bath is hotter than the sauna. Another example is copper-coated pots which are excellent conductors of heat.

Most homes have automatic thermostats that register changes in temperature. How does a change in temperature control a thermostat? Chemists as well as engineers attribute the change to metals bending toward or away from thermostat magnets because of specific heats. Respectfully, chemists use facts and data to study thermochemistry.

TERMS

- Calorie – the unit of heat necessary to raise 1 gram of water 1°C.
- Joules – newton meter or average kinetic energy (4.18 joules -1 calorie).
- Specific heat – the temperature requires to raise 1 gram of a substance 1°C.

COMMON SPECIFIC HEATS:

Note C_p (specific heats) are in cal/g°C and joules/g°C

H_2O	= 1 cal/g°C	or	4.18 joules
Pb	= 0.031 cal	or	0.1296 joules
Fe	= 0.1076 cal	or	0.449 joules
Cu	= 0.092 cal	or	0.384 joules
Al	= 0.216 cal	or	0.90 joules

THERMORCHEMISTRY PROBLEMS

Scientists can determine the amount of calories or heat needed to change the temperature of a substance by the formula:

Q = mx sp. ht. x Δt

Q = amount of heat

m = mass (grams)

t = change in temperature

Note specific heat is also written as C_p many technicians prefer to use joules in heat calculations.

EXAMPLE A

How much heat would be required to change the temperature of 35 grams of water from 15°C to 40°C?
Solution: Q = m x sp. Ht. x Δt

$$\frac{35\ \cancel{grams}}{H_2O}\ x\ \frac{1\ cal\ x\ (40°C - 15°C)}{\cancel{g}°C} = 875\ joules$$

To convert joules we simply multiply 4.18 joules x calories (875) = 3,657 joules.

EXAMPLE B **Find the heat needed to change.**

A chemist heats 65 grams of Fe form 30°C and stops at 185, °C sp. ht. of Fe is 0.1076 cal/g°C.
65 g Fe x 0.1076 cal/g°C x (185°C – 30°C) = 1084 calories

$$joules = 65\ g\ Fe\ x\ \frac{4.185}{1\ \cancel{cal}}\ x\ 1084\ \cancel{cal} = 4531\ joules$$

STATES OF MATTER REVIEWED

Earlier in out text we discussed, the three states of matter are solids, liquids, and gases. Temperature can change the state of any substance. Water is a good example, when at 0°C (freezing) water is a solid. As we increase from 0°C to 99°C water is a liquid. Finally at 100°C water is a vapor or steam. Since chemistry is a broad field, we will concentrate on some water conversion constants.

heat of vaporization – water form liquid to steam 540 cal/g°C

More specific heats: sp. ht. of ice and steam is 0.5 cal/g°C, ice to liquid = 80 cal/g°C

EXAMPLE C

How many calories will be liberated in changing 18 grams of ice at 0°C to liquid at 0°C?

Solution: mass x 80 cal/g°C = 18g x 80 cal/g°C = 1440 cal or 6019.2 joule

EXAMPLE D

How many calories will be liberated in changing 2.5 moles of water at 100°C to steam at 100°C?

Step 1 Convert moles to grams

$$2.5\ \cancel{moles}\ H_2O\ x\ \frac{18g}{1\ \cancel{mole}} = 45\ grams$$

Step 2 grams x 540 calories or heat of vaporization constant
= 45g H_2O x 540 cal/g°C = 24 300 calories

Note: kilocalories = 24 300 ~~cal~~ x $\frac{1\ Kcal}{1000\ \cancel{calories}}$ = 24.3 $Kcal$

Note: joules = 24 300 cals x 41.8 joules = 101 574 joules

Note: kilojoules = 101 574 ~~J~~ x $\frac{1\ Kilojoule}{1000\ \cancel{joules}}$ = 101.574 KJ

Note: 1.5 moles H_2O (ice) to grams = 1.5 $\cancel{moles}\ x\ \frac{18g}{1\ \cancel{mole}\ ice} = 27g\ of\ ice$

Example E

How many calories are requires to change 1.5 moles of ice at -20°C to steam at 145°C?

Solution: Several steps must be completed for the following sample problem because ice or water goes through several states before vaporization.

Example F

The states are:

1.	Ice to ice	-20°C	to	0°C
2.	Ice to liquid	0°C	to	0°C
3.	Liquid to liquid	0°C	to	100°C
4.	Liquid to steam	100°C	to	100°C
5.	Steam to steam	100°C	to	45°C

Solution: For 1.5 moles ice at -20°C to steam at 145°C convert 1.5 moles to ice = 1.5 x 18 (1 mole of water) = 27 grams

1. Ice to ice = 27g x 20 x 0.5 (sp. ht. of ice) = 270 calories
2. Ice to liquid = 27g x 80cal/g°C = 2160 calories
3. Liquid to liquid (no change of state) = 27 g x 100 x 1 = 2700 calories
4. Liquid to steam = 100°C to 100°C, P = O
   27g x 540 cal = 14580 calories
5. Steam to steam = 27g (ΔT) 45 x 0.5 = 607.5 calories

To obtain the correct answer we add all 5 steps: 270 + 2160 + 2700 + 14580 + 607.5 = 20317.5 calories

UNKOWN SPECIFIC HEATS

Chemists can determine the specific heat for a metal by experimentation. The procedure or formula is simple. The formula is Q = m sp. ht. x ΔT = cal.

What does all this mean? The heat lost by the metal is gained by the water. For example, if a metal at 90°C is placed in 5 grams of water at 20°C and the final temperature of the water is 30°C, then the final temperature is 60°C. We will now work an unknown specific heat problem.

PROBLEM

Twenty-five grams of metal at 125°C are placed in 8 grams of water at 22°C. The final temperature of the solution is 38°C. Find specific heat of the metal.

Solution: Q = m x sp. ht. x ΔT = cal

In solving for unknown specific heat we *first subtract the two water temperatures (step 1).* This will give us the number of calories. Our next step *(step 2) is important because we subtract the metal temperature from the final water temperature.* In *step 3 we find the specific heat.*

Step 1: Work with ΔT of H_2O x mass of H_2O = cal
38°C – 22°C = 16°C x 8g = 128 cal (final calories = 128)
Step 2: Metal T 125 – 38°C = 87°C
Step 3: 3Q = sp. ht. x ΔT = cal
25g (metal) sp. ht. x 87°C = 2175 calories

$$sp.ht. = \frac{128\ calories}{2175\ calories} = 0.0588\ cal/g°C$$

Proof: Q = M x sp. ht. x ΔT
25 grams x 0.0588 x 87°C = 128 calories

Summary and formula for unknown specific heat = $\frac{H_2O\ calories\ sp.ht.or\ Cp}{metal\ calories}$

mass of H_2O x ΔT H_2O = calories of H_2O

mass of metal x (metal °C – final H_2O) = $\frac{H_2O\ calories\ or\ joules}{metal\ calories\ or\ joules}$

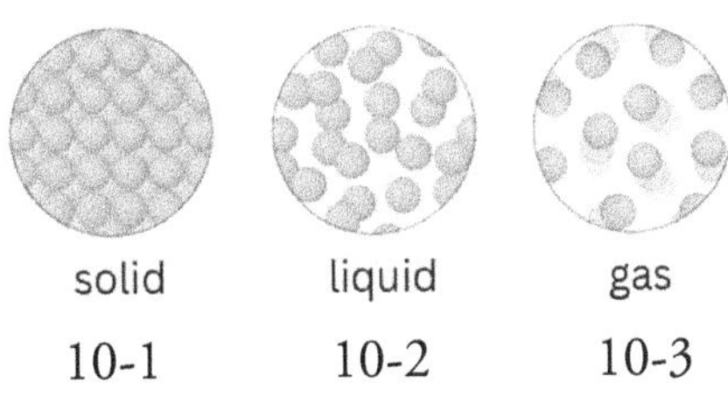

10-1 10-2 10-3

Note: Solid is 10-1: Solid (ice) — 5°C to 0°C
Note: Liquid is 10-2: At °C — note that heat will increase solubility liquid is 0°C to 100°C
Note: Gas is 10-3: At 100°C — molecules of H_2O or steam

EXERCISE 10-1

1. How many calories and joules are required to change 35 grams of Mg from 15°C to 73°C? Sp. ht. of Mg is 0.245cal/g°C.
2. 60 grams of ice at 0°C was converted to liquid at 0°C. How many calories and joules were required in the reaction?
3. How many calories are required to change 3.8 moles of water from 40°C to 56°C?
4. How many calories and joules are required to convert 24 grams of ice at -14°C to steam at 150°C?
5. How many calories are required to change 11 grams of water from 27°C to 48°C?
6. 94 grams of a metal at 60°C are placed in 125 grams of water at 36°C. the final temperature of the mixture is 46°C. what is the specific heat of this metal? Answer in calories and joules.
7. 123 grams of a metal at 75°C are placed in 250 grams of water at 25°C. the final temperature of the mixture is 40°C. what is the specific heat of this metal? Answer in calories and joules.
8. Popcorn is a favorite treat at baseball games, movies and parties and is eaten by millions. A chemist wants to know what makes popcorn pop. (Hint: Water is converted to ________) also describe the constants involved.

Refer to Diagrams (Questions 9-13)

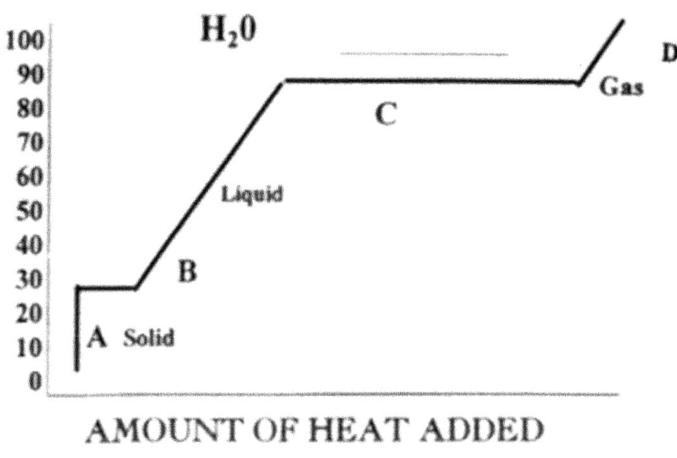

Figure 10-4

Note how ice melts as temperature increases. Observe the three states of matter.

9. The melting stage is letter ________.
10. In this phase molecules are active, far apart and freely moving around.
11. The liquid phase is ________.
12. A chemist has no change of state and his temperatures are 35°C to 95°C. This phase is ______.
13. How many calories and joules are required to change 40 grams of phase C at 100°C?
14. Convert 30 grams of ice at -12°C to steam at 160°C. answer in joules and calories.
15. Define the following terms.

- Thermochemistry
- Joule
- Calorie
- Specific heat

CHAPTER 11

SOLUTIONS

Objective and Goals

To be able to:

- solve for molarity
- compare molarity, molality and normality
- study colligative properties of some solutions
- study various types of solutions
- master solution vocabulary
- master selected boiling and freezing point problems

Keywords:

- concentration
- dilute
- liters
- milliliters
- molarity
- molality
- normality
- solute
- solvent
- solution
- saturated
- supersaturated solution
- unsaturated solution

CHAPTER 11

SOLUTIONS

A solution is a solute and a solvent retaining the properties of both. For example, soap and water is a solution because soap, the solute, is present and water, the solvent, has its properties.

TERMS

Solvent: A substance that breaks down another compound. Water is a universal solvent. Some other solvents are alcohols, acids and gasoline.

Solute: A substance broken down by the solvent. Some examples of solutes are alka seltzer, powdered soft drinks, soaps and grease.

Types of Solutions:

- Unsaturated — Excess solvent could accept more solute at the same temperature.
- Saturated — Contains all the solute it can dissolve at a certain temperature. Equilibrium is established.
- Supersaturated — Excess solute and a minimum solvent (example-crystals).
- Concentrations — Concentrations are a major concern of chemists. Concentrations tell the quantity of solvent and solute present in a solution. We can express concentrations in several ways.
- Molarity – moles per kiter or $m = \frac{moles\ of\ solute}{liters\ of\ solvent}$
- Molality – moles per Kg or 1000 grams or $\frac{moles\ of\ solute}{kilograms\ of\ solvent}$
- Normality – (N) equivalents or molarity x net positive valence of compound or

$$\frac{number\ of\ gram\ equivalent\ x\ weight\ of\ solute}{liter\ of\ solution}$$

EXERCISE 11-1

Molarity Problems [M]

Example A

A chemist dissolves 15 grams of $AgNO_3$ in 150 mL of water. What is the molarity?

Explanation for example A.

Since molarity is moles/liters of solution we factored the problem. The data concerning the solvent in the denominator. The solute data in the numerator is then converted to moles, and data in the denominator to liters.

$$conversion\ molarity = \frac{moles\ of\ solute}{liters\ of\ solvent}$$

$$\frac{15\ \cancel{grams}\ of\ AgNO_3}{150\ \cancel{mL}\ H_2O}\ x\ \frac{100\ \cancel{mL}}{1L}\ x\ \frac{1\ mole\ AgNO_3}{170\ \cancel{grams}} = 0.588\ moles = 0.588\ M$$

Example B

A technician dissolves 40 grams of Pb NO_3 in 80 mL of H_2O. Find molarity.

Solution: $\frac{40\ \cancel{grams}}{80\ \cancel{mL}}\ x\ \frac{1000\cancel{mL}}{1L}\ x\ \frac{1\ mole\ PbNO_3}{269\cancel{g}} = 1.86\ molar$

Example C

A chemist wants to prepare 150 mL of a solution of 0.5M $Al_2(SO_4)_3$. How many grams does he need?

Solution: *mass of compound* x *volume* x *molarity*

$$\frac{342\cancel{g}\ Al_2(SO_4)_3}{1\ moles}\ x\ \frac{150\cancel{mL}}{1000\cancel{mL}}\ x\ \frac{0.5M}{1M} = 25.65g\ of\ Al_2(SO_4)_3$$

Example D

How many grams are needed to prepare 1.65 liters of 0.5M solution of glucose $C_6H_{12}O_6$. The molar mass of glucose is 180 grams.

Solution: *mass* x *volume* x *molarity*

$$\frac{180\ \cancel{grams}}{1\ \cancel{moles}}\ x\ \frac{1.65\ \cancel{liters}}{1\ \cancel{liter}}\ x\ \frac{0.5\ \cancel{moles}}{1\ \cancel{liter}} = 148.5grams$$

Example E

Find molarity with 3.0 moles of NaOH in 250 mL of H_2O.

Solution: $molarity = \frac{note\ mole}{liter}$

We can factor this problem

$$\frac{3.0\ moles\ of\ NaOH}{250\cancel{mL}} x \frac{1000\cancel{mL}}{1\ liter} = 12M$$

Exercise 11-2

Molarity Problems

What is the molarity of a solution if a student dissolves 35 grams table sugar $C_{12}H_{22}O_{11}$ in 150 grams of water? Molar mass of sucrose is 342 grams.

Note: molarity is moles of solute/Kg of solvent molarity and molarity are interchangeable if water is the solvent because 1 gram of water = 1mL of H_2O.

1000mL = 1 kilogram of H_2O.

Solution: note we factor grams and moles

$$\frac{35\cancel{g}\ C_{12}H_{22}0_{11}}{150\cancel{g}\ H_20} x \frac{1000\cancel{g}}{1Kg} x \frac{1\ mole\ C_{12}H_{22}0_{11}}{342\cancel{g}} = 0.68m$$

FREEZING AND BOILING POINT PROBLEMS

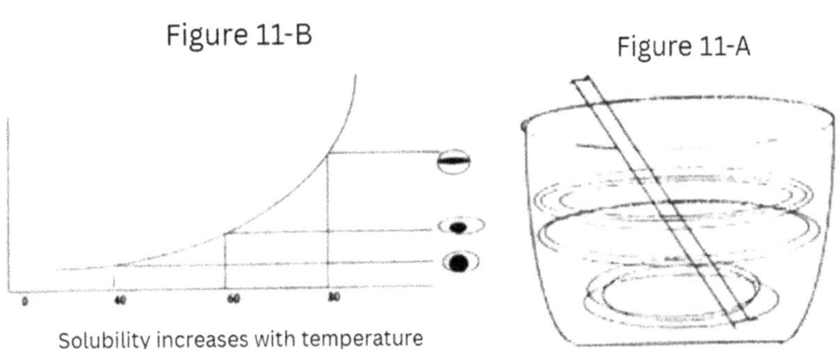

EXERCISE 11-3

Molarity Problems:

Calculate the molarity of a solution containing:

A. 1.5 moles of $C_6H_{12}O_6$ in 686cm^3 of solution
B. 64 grams of C_2H_5OH in 75mL of solution
C. 125 grams of NaOH in 3.25L of solution
D. A solution is prepared by dissolving 40 grams of $KClO_3$ in 65mL water.

Find molarity.

- What volume of this solution contains 0.4moles of $KClO_3$
- How many moles of $KClO_3$ are there in 100mL of this solution?

EXERCISE 11-4

Find molarity for A-D.

A. 3.5 moles of $PbNO_3$ in 450 grams of water
B. 60 grams of NH_4NO_3 in 170 grams of water
C. 28 grams of $AgNO_3$ in 30 grams of water
D. 125 grams of glucose in $C_6H_{12}O_6$ in 200 grams of water

Example A: What is the N (normality) of a .3M $Al_2(SO_4)_3$?
Solution: Multiply the metal or positive ion subscript by its group number. **$Al_2(SO_4)_3$ = Al 2 x 3 = 6.00**. The net positive valence is 6. Therefore, **6 x 0.3M = 1.8N**.

Example B: A chemist has a **4M** solution of H_2SO_4. Find normality. When **H** is present, we multiply **H** by its subscript, H_2SO_4which has **2** hydrogens.
Solution: H_2SO_4= 2 x 1 = 2 net positive valence = **2 x 4M = 8N H_2SO_4**
Net positive valence x N = 2 x 4M = 8N H_2SO_4

COLLIGATIVE PROPERTIED OF NON-ELECTROLYTE SOLUTIONS

The freezing points of aqueous solutions are lower than pure water. *The freezing point of water is lowered by 1.86°C per molal for a nonelectrolyte. Since pure water freezes at 0°C.*

Example a 2 molal solution of glucose (nonelectrolyte) will freeze in a water at **1.86°C** *(1 molal)* **x 2m = 0 - 3.72°C = -3.72°C**

The boiling of water for a nonelectrolyte solution will increase by 0.52°C per molal. However since water boils at 100°C we add the constant to 100°C.

Example a 1.5 molal solution of glucose will boil at **0.52°C x 1.5m = 0.78°C + 100°C = 100.78°C**

FREEZING AND BOILING POINT PROBLEMS

What is the freezing point and boiling point of a solution of antifreeze if a chemist adds 35 grams of antifreeze (ethylene glycol) to 150 grams of water. Antifreeze is $C_2H_6O_2$.

Step 1 We convert to molality using the factoring method.

$$\frac{35g\ of\ antifreeze}{150g\ H_2O} x \frac{1000g}{1Kg\ H_2O} x \frac{1\ mole\ antifreeze}{62\ grams} = 3.7m$$

Step 2 We multiply molality by constants
1 molal boiling point = 0.52°C x 3.7m = 1.92°C + 100°C = 101.92°C

Freezing point = 1.86°C x 3.7m = 6.88°C, 0°C – 6.88°C = - 6.88°C

Define the following terms:

- Miscible
- Immiscible
- Solute
- Solvent
- Solubility
- Saturated solution
- Unsaturated solution
- Colligative properties
- Molarity
- Molality
- Normality
- Part per million
- Millimole
- Osmosis
- Diffusion
- Supersaturated solution

EXERCISE 11-5

Solve the normality or molarity.

3M H_2SO_4 ________ N
12N H_2SO_4 ________ N
1.5M HCl ________ N

A chemist adds 60 grams of ($Al_2(SO_4)_3$) to 500 grams of water. What is the molarity? What is the normality of this situation?

Nonelectrolyte (boiling and freezing point problems) in water.

A. 125 grams of $HgNO_3$ is added to 300 grams of water. What is the b.p. and f.p. of this solution?
B. What is the b.p, and f.p. of a solution containing 25 grams of glycerin (M.W. 92) per 250 grams of water?
C. What is the molecular weight of a substance if 63 grams of it dissolved in 200 grams of water and makes a solution which boils at 101.04°C?

FOR FURTHER STUDY

1. Dr. Liz recorder a b.p. of 102.04°C for patient "X" glucose (dextrose). Complete the following:

Molality	________
Grams in 0.5 liters	________
Freezing point	________
Grams of glucose in 0.25 liters	________
Grams of glucose used to prepare 1 liter	________

2. What volume of 1.20M $HC_2H_3O_2$ (vinegar) solution contains .30 moles of vinegar?
3. An electrolyte differs from a non-electrolyte because each ion has molal properties. For example, **$CaCl_2$→2Cl = Ca** or **3** ions. Since each ion = 1 molal, what is the **f.p.** and **b.p. of $CaCl_2$** in water for the following?

	f.p.	b.p.
1 molal	________	________
2 molal	________	________
0.65 molal	________	________
Solve for electrolyte NaCl→Na→Cl		
2 molal	________	________
0.75 molal	________	________
1.0 molal	________	________

Molarity Problems

A technician has the following data for patient X.

7.74×10^{-3} mole of HCl

5.0×10^{-2} liters of solution

The doctor must know the acid concentration of patient X's stomach. What is the M or stomach acid concentration?

Molarity

A. Describe how you would prepare 550mL of a **0.45M solution of** $\mathbf{Ca(OH)_2}$?

B. A solution has 28 grams of sucrose $\mathbf{C_{12}H_{22}O_{11}}$ in 335mL of solution. Find the molarity of this solution.

C. What is the volume if 12g of KOH are dissolved to form a **0.65M solution**?

D. How many moles are there in 300mL of: a **0.275M solution**?

E. A solution was made by dissolving 20g of $\mathbf{HNO_3}$ in 250mL of solution.

Calculate:

a. The molarity of this solution

b. The volume of solutinused to prepare .20 moles

c. The moles in 500mL of solution

CHAPTER 12

Titrations

Objectives and Goals

To be able to:

- define titration
- study its lab applications
- determine the strength or concentration of acids and bases
- study neutralization reactions

Keywords:

- acids
- bases
- indicators
- equivalent weight
- titration
- dilution
- molarity
- normality
- neutralization
- salt
- burettes
- litmus paper
- pH paper

CHAPTER 12

TITRATIONS

Titrations is a laboratory procedure used to determine the concentration of an acid or base. This procedure involves the use of a burette and an indicator. As stated earlier, red is an acid and blue is a base.

Titrations involved the reaction of an acid solution with a base solution. In lab work, the chemist knows the volume and concentration of the acid or basic solution. The reaction of an acid and base is a neutralization reaction.

acid + base → salt + water

Exercise 12-1

Write an essay on uses of titrations

Titrations Problems

If 40mL of a .600m solution of NaOH is requires to neutralize 80ml of an H_2SO_4 solution, what is the molarity of the H_2SO_4 solution?

Balance equation: $2NaOH + H_2SO_4 – Na_2SO_4 + 2H_2O$
Note: 2 moles of NaOH reacts with 1 mole of H_2SO_4

Solution: $$\frac{ml\ x\ Molarity\ of\ NaOH}{ml\ x\ H_2SO_4\ x\ moles\ of\ NaOH}\ x\ moles\ of\ H_2SO_4$$

$$\frac{40\cancel{ml}\ x\ 0.6m\ NaOH}{80\cancel{ml}\ x\ H_2SO_4\ x\ 2}\ x\ \frac{24\cancel{ml}}{160\cancel{ml}} = 0.15M$$

Proof: $$\frac{0.15M\ H_2SO_4\ x\ 80\cancel{ml}\ x\ 2}{40\cancel{ml}} = 0.6M$$

What volume in ml of 0.700m HNO_3 is required to neutralize 55ml of a 0.4m NaOH
HNO3 + NaOH→NaNO3 + H2O

Note: 1 mole reacts with 1 mole

$$\frac{55ml\ of\ 0.4\ NaOH}{0.700m\ HNO_3} = 31.43\ ml\ HNO_3$$

How much water must be added to 85ml of a 0.6M NaOH solution to make a 0.4M solution?

Solution: $$\frac{85ml\ x\ 0.6M\ naOH}{0.4M} = 127.5ml\ of\ water$$

Exercise 12-1

1. 50.1 ml of a 0.2N HCl is required to neutralize 40.0 ml of an unknown base. Find the molarity and normality of the base.
2. How much water must be added to 70 ml of a 0.25M HNO_3 to make a 0.18M solution?
3. A 35 ml sample of KOH reacts completely with 20 ml of a 0.2N H_2SO_4. Find the molarity of KOH.
4. How many ml of 0.30N KOH required to titrate 40 ml of a 6.0 HNO_3?
5. What is the molarity of a stock solution of HCl if a student diluted 52 ml of HCl to prepare 160 ml of a 0.7M solution?
6. How many ml of a 1.5M solution X are needed to obtain 0.8 moles (hint: moles/molarity) x 1000 ml?
7. What is the molarity of a solution if a chemist diluted 60 ml of a 0.3M HCl to 140 ml?
8. How many moles are there in 350 ml of a 9.7M solution?
9. How would you prepare 400 ml of a 0.85M HNO_3 from a 6.0M HNO_3 solution?
10. A student has 60 ml of 0.8M HCl. He used 20 ml of NaOH to neutralize the HCl. What is the molarity of NaOH?

Note: The equivalent weight of a compound is the net positive valence divided by its formula weight.

Example H_3PO_4 subscript of Hydrogen x Group = 3 x 1 = positive valence
Equivalent weight of H_3PO_4 = 98g (1 mole of H_3PO_4) ÷ 3
98 ÷ 3 = 32.66 grams
Na_2SO_4 1 mole = 142.1g
Positive valence = 2 (Na + 2 x 1)
Equivalent = 142.1 ÷ **2 = 71.05g**

FOR FURTHER STUDY

A Few Commercial Uses of Titration

Generic ammonia vs. consumers choice ammonia

A customer buys generic ammonia (NH_3) because the price is lower. After switching a television commercial her curiosity convinced her to compare the strength of a name brand and her generic brand. She gathered the following data:

20 ml of 0.5 HCl was used to neutralized 85 ml of generic NH_3. What is the M of this generic ammonia?

40 ml x 0.25M HCl was used to neutralized 85 ml of name brand NH_3. What is the M of this ammonia? Analyze your results (list and compare)

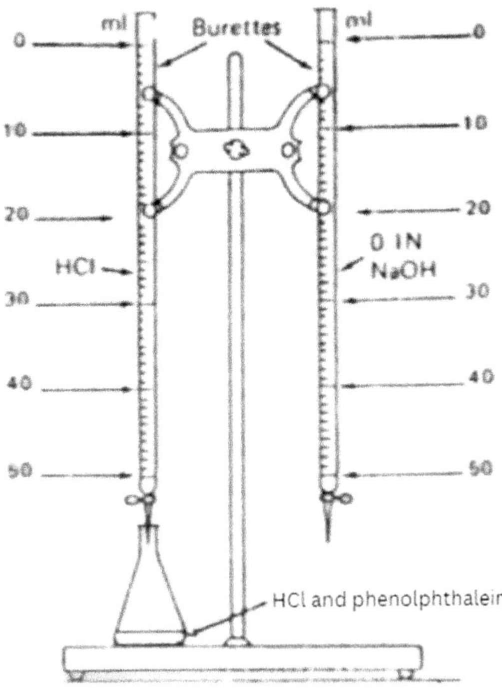

Figure 12-1-A

Titrate the following:

Vinegar C 60 ml of 0.25M NH_3 titrated 40 ml of vinegar
Vinegar D 60 ml of 0.30M NH_3 titrated 40 ml of vinegar
Analyze your results, list molarities and compare.

Chemists often use indicators when doing titrations. One common acid and base indicator is red cabbage juice. Acids turn red and bases turn green or blue because H_2 reacts with the cabbage.

Furthermore, titration is easily observed when acids change from red → clear or red → green. Likewise, bases change from greenish blue → light red. We must emphasize there is a color change in titrations based on the indicator used.

FOR FURTHER STUDY

To titrate an acid use a base with a known molarity and record volume of the base. This is added to a known volume of acid.

For acid: $$\frac{M\ of\ base\ x\ volume\ of\ base}{volume\ of\ acid} = M$$

For base: $$\frac{M\ of\ base\ x\ volume\ of\ acid}{volume\ of\ base} = M$$

Figure 12-1-B

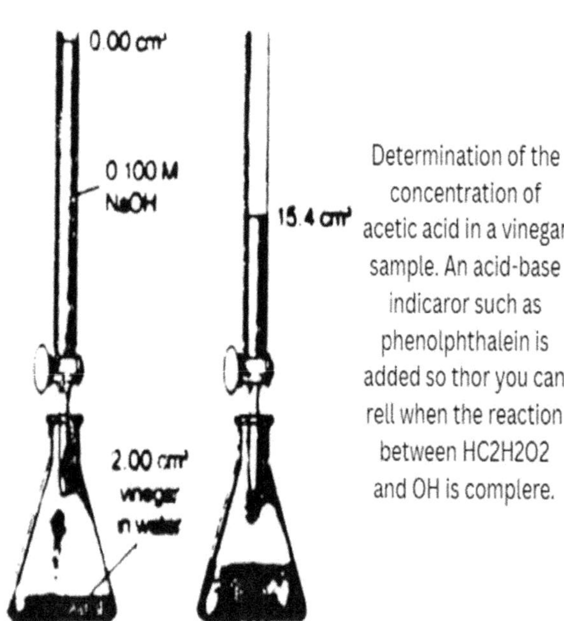

Determination of the concentration of acidic in a vinegar sample. An acid-case indicator such as phenolphthalein is added so that you can tell when the reaction between $HC_2H_3O_2$ and OH^- is complete.

CHAPTER 13

Redox Equations

Goals and Objectives

To be able to:

- balance redox equations
- identify oxidizing agents
- identify reducing agents
- use algebra to balance redox equations

Key words:

- oxidation
- reduction
- equations
- oxidation number
- half cell
- half equations
- redox equations

CHAPTER 13

REDOX EQUATIONS

OCIDATION AND REDUCTION

General information:

Oxidation – is defined as the loss of electrons.
Reduction – is defined as the gain of electrons.
Oxidation is an element that increase its oxidation number and is oxidized.
Example: $Pb \rightarrow Pb^{++} + 2e$- Pb is increased by 2
Reduction is an element that decreases its oxidation number and is reduces.
Example: $Cl_{\circ 2} + 2e \rightarrow 2Cl$
The above example shows chlorine being reduced from 0 to -1.
Reducing Agents lose electrons and the final result is oxidations.
Oxidizing Agents gain electrons and the final result is reduced.
Redox equations are equations that involve a change in oxidation number. If the oxidation numbers remain the same, it is not a redox equation.

RULES FOR BALAMCING REDOX EQUATIONS

1. Calculate the oxidation number of each compound.
2. Be sure the oxidation number is divided by the element's subscript. O_2 is -2 not -4 because -2 x +2 = -4, but we subscript 3 or -4 ÷ 2 = -2
3. An element in the free state of uncombined has an oxidation number is zero.
4. Water should be balanced last.
5. Refer to Special Topics oxidation number.

OXIDATION REDUCTION REACTIONS

Some chemical reactions are like telephone. In a telephone call, we have and exchange of words with one or more people. Likewise, in oxidation the reduction reactions we have an exchange of electrons or + and – charges transferred. Several methods are commonly used to balance redox equations.

BALANCING REDOX EQUATIONS

Notes: oxygen is -2 unless in a peroxide when it is -1
hydrogen is +1 unless in a hydride when it is a -1
A student must remember the oxidation numbers, oxygen and hydrogen. Also remember that the group 1 elements have a +1 oxidation number and most of group 7 are -1.

BALANCE USING THE OXIDATION STATE METHOD

Determine the two elements which have caused oxidation states in the reaction.

O   +1 – 1 = 0   +2 -2   0 Mg from 0 to +2 oxidized H from $^{+}1$ to 0 reduced
$Mg + HCl \rightarrow MgCl_2 + H_2$ Mg loses $2e^-$
$Mg^0 \rightarrow Mg^{+2}$ $H+ \rightarrow \mathbf{H^o}$ H gains $1e^-$

Since redox equations emphasize exchange of electrons, we multiply the elements by the opposite electrons.

Mg (oxidized) lost $2e^-$ So H is 2 x 1 or H2
H+ (reduced) gained $1e^-$ So Mg is 1 x 1 or Mg
Balanced: $Mg + 2HCl \rightarrow MgCl_2 + H_2$ gas

Solution 2 Half Reduction Method

Half equation ion electron method or method (often aqueous solutions) involved 4 steps:

1. Split the equation into oxidation and reduction.
2. Balance one of the half equations with equal of atoms.
3. Balance the other half of the equation.
4. Combine the two half equations in a way to eliminate electrons.

EXAMPLE $MnO^-_{4\,(aq)} + Fe^{+2}_{(aq)} \rightarrow Mn^{+2}_{(aq)} + Fe^{+3}_{(aq)}$

Step 1: Oxidized $Fe^{+2}\,_{(aq)} \rightarrow Fe^{+3}\,_{(aq)}$ (2a)
by 1   +7-8
Reduced $MnO_{4\,(aq)} \rightarrow Mn_{+2\,(aq)}$
by 5

Step 2: Half equation 2a is balanced
$Fe^{+2}\,_{(aq)} \rightarrow Fe^{+3}\,_{(aq)}\,^{+e-}$

Step 3: Balance other half of equation
Note: Check atoms on both sides such as $MnO_{4\,(aq)} + Mn^{+2}_{(aq)} \rightarrow Mn^{+2}_{(aq)} + 4H_2O$
Note: $4H_2O$ balance the four oxygen atoms in MnO_4^-
To complete this half we need $8H^+$ due to $4H_2O$
$MnO^-_{4\,(aq)} + 8H^+_{(aq)} \rightarrow Mn_{(aq)}\,^{2+} + 4H_2O$

Step 4: We now multiply by opposite electron charges
Fe x 5 (Mn ↕ 5) MnO_4^- by 1 (Fe ↕ 1)

BALANCED EQUATION

$MnO_{4\ (aq)}^{-} + 8H^{+}_{(aq)} + 5\ Fe^{2+}\ Mn_{(aq)}^{2+} + 4H_2O + 5\ Fe_{(aq)}^{3+}$

BALANCED USING THE OXIDATION STATE METHOD

Figure 13-A

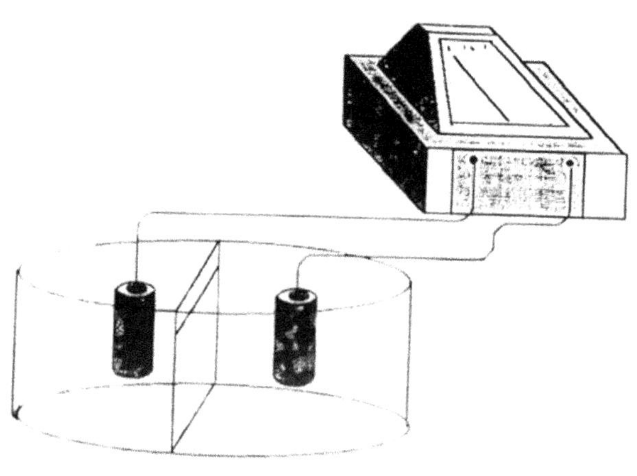

Half-cell reaction – positive and negative charges

EXERCISE 13-1

Balance the following redox equations, list oxidizing and reducing agents. Also list substance oxidized and reduced.

1. $CdS + I_2 + HCl \rightarrow CdCl_2 + HI + S$
2. $SnCl_2 + HgCl_2 \rightarrow SnCl_4 + HgCl$
3. $NaC10 + H_2S \rightarrow NaCl + H_2SO_4$
4. $Ag + HNO_3 \rightarrow AgNO_3 + NO + H_2O$

EXERCISE 13-2

1. $Zn + HCl \rightarrow ZnCl_2 + H_2$
2. $Cu + AgNO_3 \rightarrow Ag + Cu(NO_3)_2$
3. $HNO_2 + HI \rightarrow NO + I_2 + H_2O$
4. $K_2CO_3 + Br_2 \rightarrow KBr + KBrO_3 + CO_2$
5. $K_2Cr_2O_7 + HCl \rightarrow KCl + CrCl_3 + H_2O + Cl_2$
6. $MnO_2 + HCl \rightarrow MnCl_2 + H_2O + Cl_2$
7. $PbO_2 + HI \rightarrow PbI_2 + I_2 + H_2O$
8. $NH_3 + O_2 \rightarrow NO + H_2O$
9. $MnO_4^{2-} + Cl_2 \rightarrow MnO_{4-} + Cl^-$

CHAPTER 14

Electrochemistry

Goals and objectives

To be able to:

- study electricity and chemical reactions
- study electrolysis
- study standard reduction potentials
- calculate voltage
- predict chemical reactions
- use a standard reduction potentials chart
- study Faraday's laws

Keywords:

- volts
- amperes
- ohms
- cations
- anions
- anode
- cathode
- Faraday
- Coulombs

CHAPTER 14

ELECTROCHEMISTRY

Electrochemistry is defined as the study of electricity in chemical reactions. In this brief chapter we will answer two questions in relation to electrochemistry.

1. What effects do electrical currents have on solutions of electrolytes.
2. What electrical effects accompany redox equations and electromotive forces (EMF).

TERMS

- Volt – unit of force
- Coulomb – quantity of electricity that will deposit 0.001118 grams of Ag in one second
- Amperes – current rate flow of one coulomb per second amps = volts / ohms
- Ohm – a unit of electrical resistance
- Cations – positive charge ions
- Anions – negative charge ions
- Cathode – negative pole of battery, re duction occurs
- Anode – positive pole of battery, oxidation occurs
- Faraday – 1 mol (mole) of electron is 96,500 coulombs give 108g of silver (Ag)

ELECTROLYSIS

Let us consider a case of electrolysis. An electric current is passed through a solution of HCl.

$HCl = H^+ = Cl^-$

A change in charge indicates oxidation and reduction.

$H^+ \ 1e^- \rightarrow H$ reduction at cathode

$Cl^- \rightarrow Cl - 1e^-$ oxidation at anode

The result of electrolysis is decomposition.

$2HCl = H^2 + Cl^2$

Selected Faraday's Problems

Constants 96,500 Coulombs = 1 Faraday, 3600 sec = 1 hour

How much Al will be deposited using $Al_2(SO_4)_3$ in 1.5 hours and 40 amps. Note Al deposited at the battery cathode.

Solution $\frac{Amps\ x\ seconds}{96{,}500\ Coulombs}\ x\ eq.wt.$

$$Al = \frac{27g}{1\ mole\ Al}\ x\ \frac{1\ mole\ Al}{3e^-} = 9g\ Al\ (equivalent\ weight\ of\ Al)$$

Note: 3e- is from group 3A or 13 A

Note: eq. wt. = $\frac{1\ mole}{electrons\ charge\ of\ the\ element}$

Solution $\frac{40\ amps\ x\ 5400\ sec}{96{,}500\ Coulombs}\ x\ 9g = 20\ grams\ or\ 20.14\ grams$

We converted 1.5 hours into seconds = 5400 seconds.

FOR FURTHER SAMPLE

Given: 0.7461 moles of Al $\frac{20.14\ g}{2}\ x\ \frac{1\ mole}{27\ g}$

3e

96 500 Coulombs

40 amps

54 000 seconds

$$seconds = \frac{Coulombs}{amps} = \frac{\cancel{96500}\ x\ 3e\ x\ 0.746\ \cancel{moles\ Al}}{40\ amps} = 5399 amps/s$$

$$amps = \frac{Coulombs}{sec} = \frac{215\ 967\ Coulumbs}{5\ 400\ sec} = 39.99$$

$$Coulombs = \frac{amps\ x\ sec}{moles\ x\ e^-} = \frac{40\ amps\ x\ 5\ 400}{3\ e\ x\ 0.746\ moles} = \frac{216\ 000}{2.238} = 96\ 514\ Coulumbs$$

$$Seconds = \frac{Coulombs}{amps}\left(coulombs = moles\ x\ electrons\ x\ \frac{96{,}500\ Coulombs}{1\ Faraday}\right)$$

$$Al = 75g\ x\ \frac{1\ mole}{27g\ Al} = 2.77\ moles\ x\ 3e^-\ x\ \frac{96{,}500\ Coulombs}{1\ Faraday} = 804166.7\ Coulombs$$

$$Al = \frac{75}{27} = 2.77\ moles\ x\ 3e^-\ x\ 96{,}500 = 804166.7\ Coulombs$$

$$Since\ seconds = \frac{Coulombs}{amps} = \frac{8041667.7\ Coulombs}{125\ amps} = 6433\ seconds$$

$$minutes = 6433\ seconds\ x\ \frac{1\ minute}{60\ seconds} = 107.2\ minutes$$

$$hours = 107.2\ minutes\ x\ \frac{1\ hour}{60\ minutes} = 1\ hour\ and\ 47\ minutes$$

ELECTROCHEMISTRY AND REDOX

TABLE OF STANDARD REDUCTOIN POTENTIALS

(Data is for substances dissolved in water in which ion concentrations are 1 molar, temperature is 298 K, and gas pressure is 1 atmosphere)

STANDARD ELECTRODE POTENTIALS		
Ionic Concentrations 1 M Water at 298 K, 1 atm		
Half-Reaction Oxidizing Agents	 Reducing Agents	E°
°2(g) + 2e^-	→ 2F^-	+2.87
MnO_4^- + 8H^+ + 5e^-	→ Mn^{2+} + 4H_2O	+1.52
Au^{2+} + 3e^-	→ Au(s)	+1.50
Cl_2(g) + 2e^-	→ 2Cl^-	+1.36
$Cr_2O_7^{2-}$ + I4H^+ + 6e^-	→ 2Cr^{3+} + 7H_2O	+1.33
MnO_2(s) + 4H^+ + 2e^-	→ Mn^{2+} + 2H_2O	+1.23
O_2(g) 2H^+ +2e^-	→ H_2O	+1.23
Br_2(l) + 2e^-	→ 2Br^-	+1.08
NO_3^- + 4H^+ + 3e^-	→ NO(g) + 2H_2O	+0.96
Ag^+ + e^-	→ Ag(s)	+0.80
Hg_2^{2+} + e^-	→ Hg(l)	+0.79
Hg^{2+} + 2e^-	→ Hg(l)	+0.78
NO_3^- + 2H^+ + e^-	→ NO_2(g) + H_2O	+0.78
Fe^{3+} + e^-	→ Fe^{2+}	+0.77
I_2(s) + 2e^-	→ 2I^-	+0.53
Cu^+ + e^-	→ Cu(s)	+0.52
Cu^{2+} +2e^-	→ Cu(s)	+0.34
SO_4^{2-} + 4H^+ + 2e^-	→ SO_2(g) + 2H_2O	+0.17
Sn^{4+} + 2e^-	→ Sn^{2+}	+0.15
2H^+ + 2e^-	→ H_2(g)	0.00

$Pb^{2+} + 2e^-$	$\rightarrow Pb(s)$	-0.13
$Sn^{2+} + 2e^-$	$\rightarrow Sn(s)$	-0.14
$Ni^{2+} + 2e^-$	$\rightarrow Ni(s)$	-0.25
$Co^{2+} + 2e^-$	$\rightarrow Co(s)$	-0.28
$2H^+ (10^{-7} M) + 2e^-$	$\rightarrow H2(g)$	-0.41
$Fe^{2+} + 2e^-$	$\rightarrow Fe(s)$	-0.44
$Cr^{3+} + 3e^-$	$\rightarrow Cr(s)$	-0.74
$Zn^{2+} + 2e^-$	$\rightarrow Zn(s)$	-0.76
$2H_2O + 2e^-$	$\rightarrow 2OH^- + H_2(g)$	-0.83
$Mn^{2+} + 2e^-$	$\rightarrow Mn(s)$	-1.18
$Al^{3+} + 3e^-$	$\rightarrow Al(s)$	-1.66
$Mg^{2+} + 2e^-$	$\rightarrow Mg(s)$	-2.37
$Na^+ + e^-$	$\rightarrow Na(s)$	-2.71
$Ca^{2+} + 2e^-$	$\rightarrow Ca(s)$	-2.87
$Sr^{2+} + 2e^-$	$\rightarrow Sr(s)$	-2.89
$Ba^{2+} + 2e^-$	$\rightarrow Ba(s)$	-2.90
$Cs^+ + e-$	$\rightarrow Cs(s)$	-2.92
$K^+ + e-$	$\rightarrow K(s)$	-2.92
$Rb^+ + e-$	$\rightarrow Rb(s)$	-2.93
$Li^+ + e-$	$\rightarrow Li(s)$	-3.00

We already stated in redox reactions an electron is transferred from a substance being reduced to a substance oxidized. Chemists can easily separate cells (beakers). A salt bridge will allow ions to move from one solution to another. The flow of electrons are transferred after the substance being reduced is connected to the substance oxidized. A charge is produced because of the continuous flow of ions through the salt bridge. This entire system is called a chemical cell. The cells convert chemical to electrical energy.

Figure 14-A

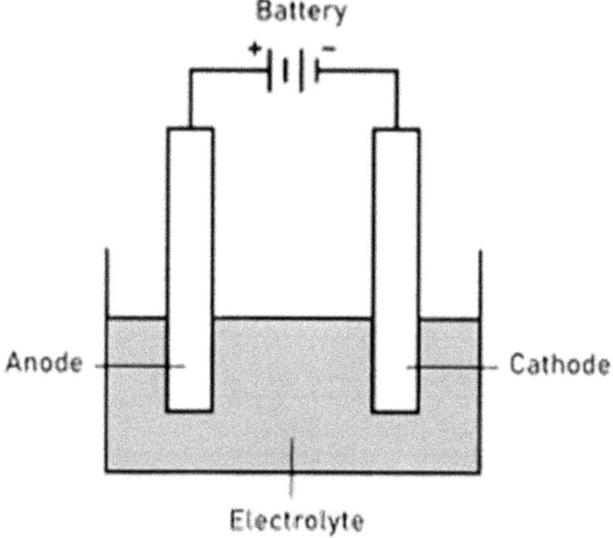

Cathode (-) Anode (+)

Metals are deposited at the cathodes. Note that cathode is negative during electrolysis.

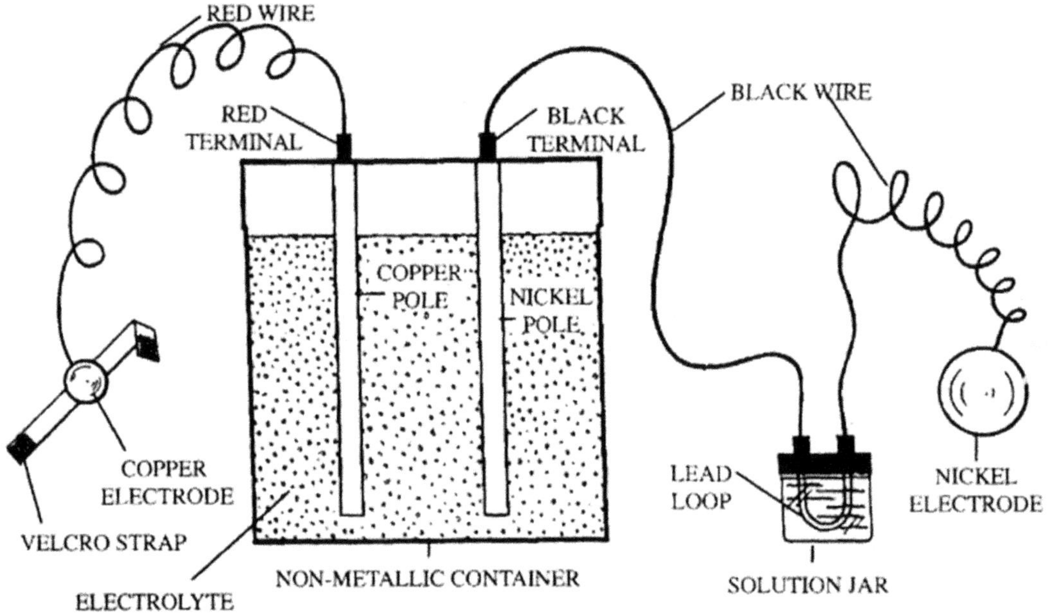

FIGURE 1: WET CELL BATTERY (SHORT-POLE TYPE)

Figure 14-B

Electronic Potential

What causes electrons to flow in a chemical cell? Electrons flow because of a difference in electrical voltage. We call this difference of electric potential. This potential, also called volt, measure the free energy of the redox reaction.

Electrode Potential

Hydrogen half reaction 2H+ + H2(g) is the standard to compare the electrode potentials of half reactions. The voltage between any given half reaction and the hydrogen half reaction is the standard electrode potential.

Net Potential

The net potential of two half cells can be determined by adding the half-cell reactions using algebra.

Rules for calculating net potential

1. Refer to a standard electrode potential chart (see appendix).
2. Find the half reaction in the chart and write its voltage as listed.
3. Since most charts are reduction potentials reverse the sign of the oxidized half cell reaction.

FARADAY'S LAW

Countless people admire gold chains, silver-plated jewelry, brass buckles, copper-coated keys, tie pins and shiny automobile bumpers. Many of these attractive items are a result of electroplating or Faraday's laws.

Figure 14-C

A Faraday's Apparatus's

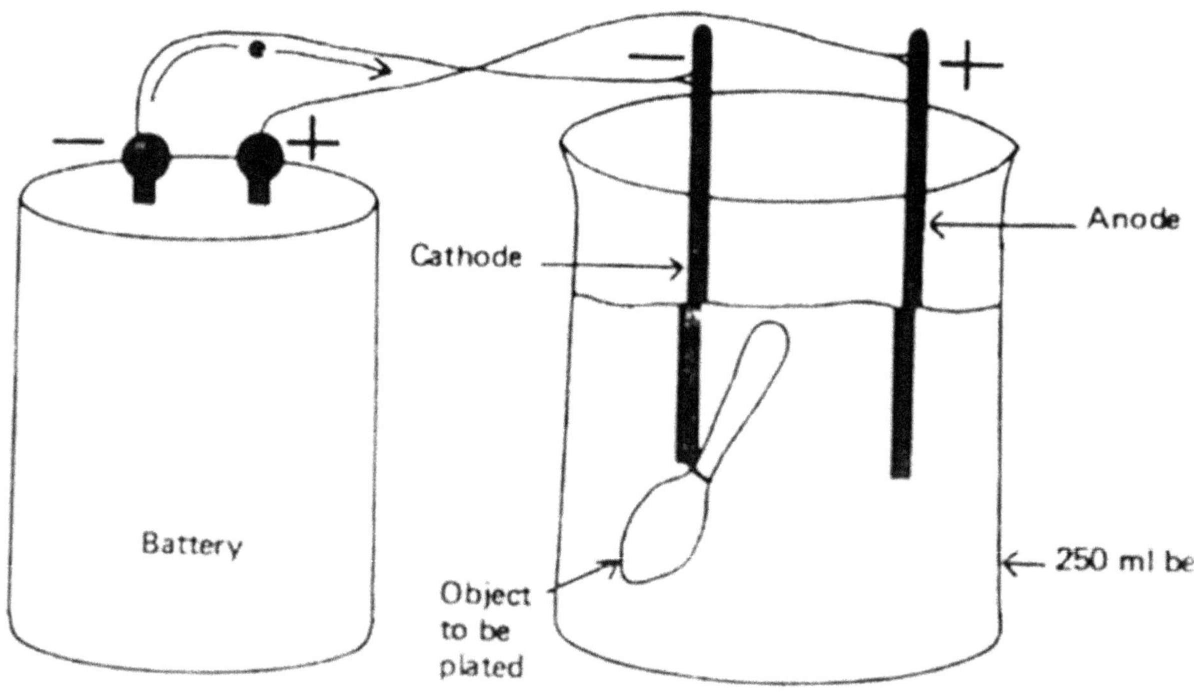

EXAMPLE 1

Find voltage $Ca + Cl_2 = CaCl_2$
First we write half cell and record voltage from standard reduction chart.

$Ca \rightarrow Ca^{2+} 2e^-$ -2.87 volts
$Cl_2 + 2e \rightarrow 2Cl^-$ +1.39 volts

Since Ca is oxidized from Ca° to +2 we must reverse the sign because it is a standard reduction potential chart.

$Ca \rightarrow Ca^{2+} + 2e^- = +2.87$
$Cl_2 + 2e^- \rightarrow 2Cl = \underline{+1.39}$
+4.26 volts

We should note that a + voltage indicated a favorable reaction. A negative potential indicated the reaction will not occur.

Chemical shorthand for above: $CalCa^{2+}/Cl_2^{\circ}/2Cl-$
Oxidized reduced

EXAMPLE 2

What is the voltage for: $Ni + Cu^{2+} \rightarrow Ni^{2+} + Cu$
Solution: (get volts from chart)

$Ni \rightarrow Ni^{2+} + 2e^-$ -0.25 volts
$Cu^{2+} + 2e^- \rightarrow Cu$ +0.34 volts

Since Ni is oxidized we must change he sign to +0.25 volts.

+0.25 Ni/Ni^{2+} Note shorthand
$\underline{+0.34}$ Cu^{2+}/Cu° $Ni/Ni^{2+}/Cu^{2+}/Cu^{\circ}$
+0.59 volts oxidized reduced

EXERCIS 14-1

1. What are some industrial uses of electrochemistry?
2. Define electrochemistry, electroplating, electric potential, coulomb, salt bridge, cathode, anode, standard reduction chart.
3. How many grams of Ca will be plated from a $CaCl_2$ solution with a current of 45 amperes flowing for 3 hours?
4. How much Cu^{+2} will be plated in 5 hours using 55 amps?
5. How long will it take to plate 88 grams of Al from $Al_2(SO_4)_3$ using current of 17 amps?
6. Calculate E° (net potential) for 6-9.
   $2\ Cr + 3\ Cu^{2+} \rightarrow 3\ Cu + 2\ Cr^{3+}$
7. $Mg + 2\ Ag^{+} = 2\ Ag^{0} + Mg^{2+}$
8. $Fe + Pb^{2+} = Fe^{2+} + Pb$
9. $Mg + 2H^{+} = Mg^{2+} + H_2$
10. A technician plated 13 grams of Cu^{2+} in 30 minutes. How many amps did he use?

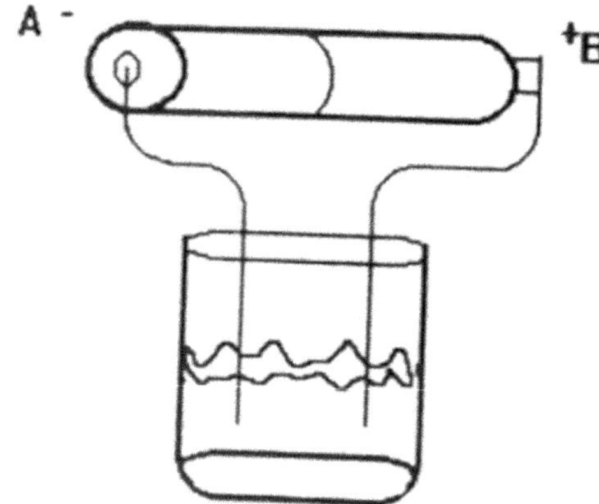

11. Define the process occurring at A and B (oxidation, reduction, etc.)
12. Letter A is ___________, letter B is ___________.
13. What letter is the best for electroplates?

CHAPTER 15

Gas Laws

Goals and objective

To be able to:

- study the laws of Boyle's, Charles, Dalton, graham and Avogadro
- solve gas law problems
- study the molar volume constant 22.4 L = 1 mole of a gas

Keywords:

- gas
- pressure
- volume
- kinetic molecular theory
- matter
- kelvins
- standard conditions or STP
- diffusion or effusion

CHAPTER 15

GAS LAWS AND THE KINETIC MOLECULAR THEORY

Kinetic Molecular Theory

A. The particles of gases are in random and rapid motion.
B. The average kinetic energy of gas is proportional to the absolute or Kelvin temperature.
C. Gas particle collisions are elastic or bounce back without any loss of energy.
D. Matter is composed of atoms and molecules which are far apart in gases.

MEASUREMENTS

Gases are measured according to Kelvins and barometric pressure. Common barometric or pressure units 760 torr (after Torricelli who invented the barometer).

The following are other pressures used by chemists to indicate 1 atmosphere at sea level:

760 mm Hg	29.92 in Hg
1 atm	14.7 lb/in^2 or psi
760 torr	101.3 kilopascals (kPa)

BOYLE'S LAW

The volume of a gas is inversely proportional to the pressure at constant temperature.

Simplified: When you increase pressure you decrease the volume of a gas.

Example People open windows when they smell cooking gas because the opened windows increase air pressure, therefore, decreasing the volume of stove gas.

Simplified: When decrease pressure you increase the volume of gas.

Example A health club steam bath is hot while the door is closed because the closed door decreases pressure resulting in an increase of steam.

CHARLE'S LAW

Charle's Law is a direct statement on gases and temperature. He stated if you *increase temperature*, the *volume of a gas will increase*. For example, technician increased the temperature of a gas flask for 30°C to 60°C resulting in more gas. On the other hand, Charles stated if you *decrease temperature*, the volume of a *gas will decrease*. Example: A cook decrease steam by turning the heat down on the stove.

Gas Law Diagrams
Figure 15-1

Gas Law Diagrams
Figure 15-1

CO_2 and Balloons

CO_2 gas at 19°C or 292 K - A

24°C or 297 K - B

33°C or 306 K - C

Note: Charle's Law says increase in temperature equal increase in volume.

Figure 15-1-A

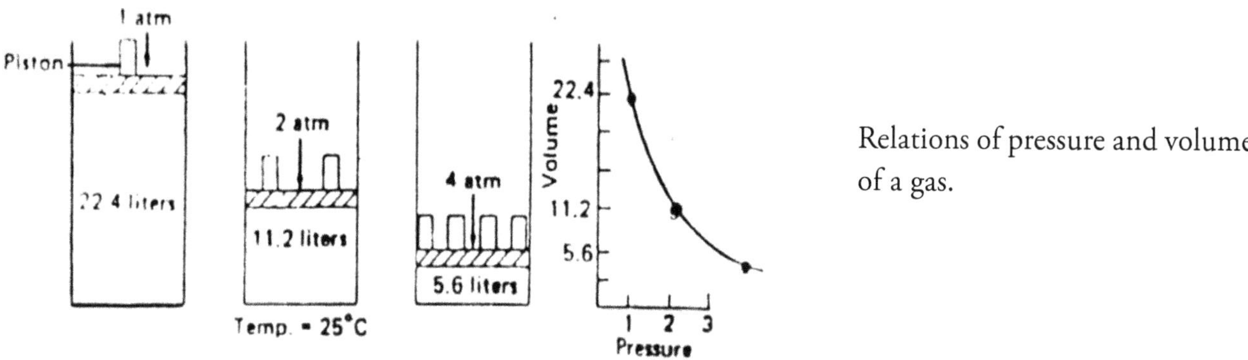

Relations of pressure and volume of a gas.

Relations of pressure and volume of a gas.

Boyles Law: Decrease pressure = ↑ volume, Increase pressure = ↓ volume

Figure 15-2
Note Torricelli's barometer

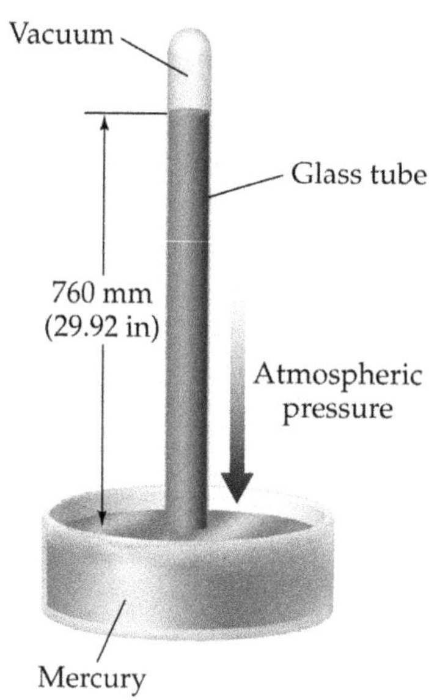

Figure 15-3
Hydrogen and oxygen molecule join to form water

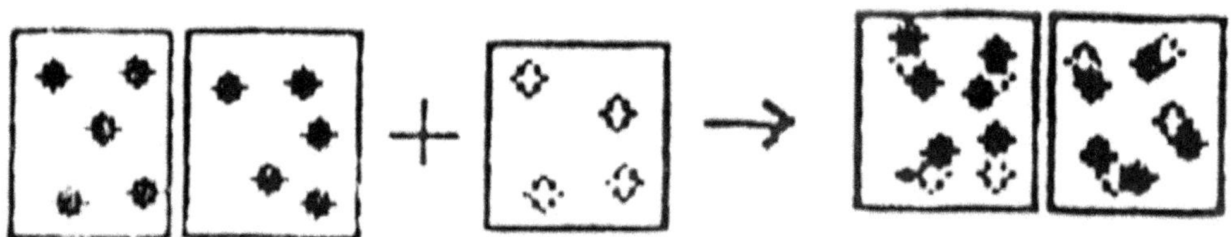

Note the formation of water. Each block is one mole so $2H_2 + O2 \rightarrow 2$ *moles of* H_2O.

Figure 15-4

Diagrammatic representation of a common method for collecting a gas.

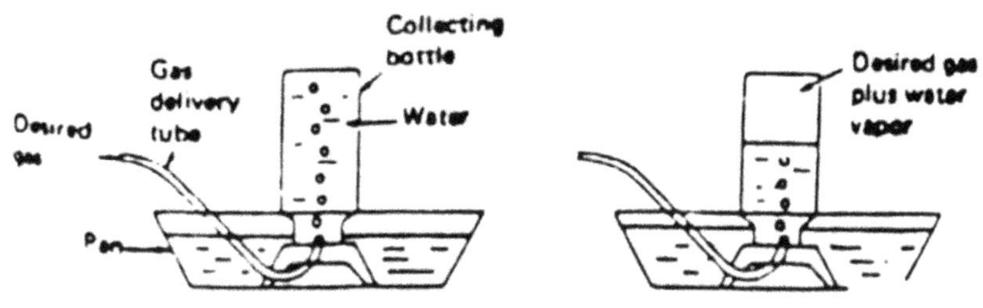

Figure 15-5

Diagrams 15-4 and 15-6 illustrate the diatomic properties of most gases.

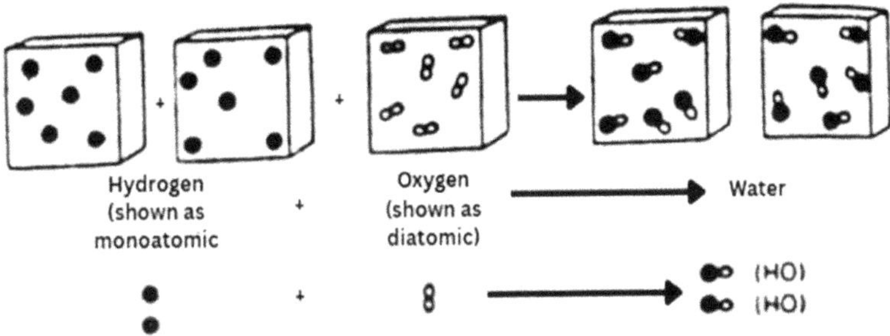

Convert 8 grams of O_2 to volume. Note 22.4 liters = 1 mole of gas.

$$8g\ O_2\ x\ \frac{1\ mole}{32\ g}\ x\ \frac{22.4\ liter}{1\ mole} = 5.6\ liters\ or\ 5.6\ dm^3$$

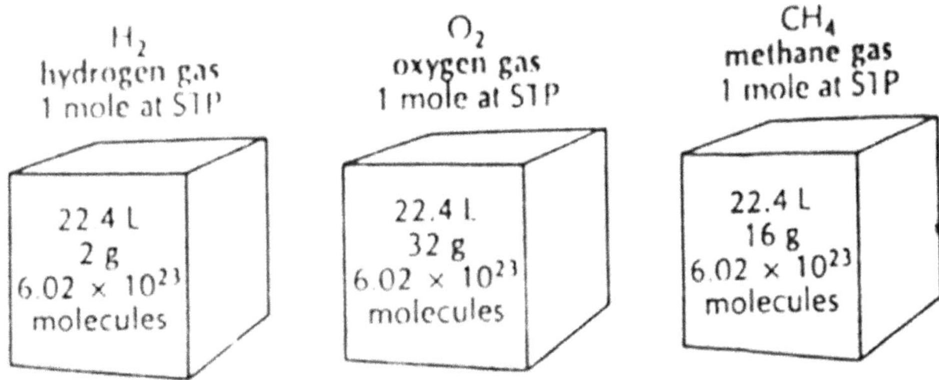

PROBLEMS SET 1

BOYLE'S LAW

What volume will 500 mL of O_2 occupy if pressure is decreased from 1520 torr to 760 torr?

Note V_1 = original volume and V_2 = new volume

Solution P_1 = 1520 torr, P_2 = 760 torr, V_1 = 500mL

$$V_2 = \frac{V_1 \, x \, P_1}{P_2} = \frac{500\ mL\ x\ 1520\ \cancel{torr}}{760\ \cancel{torr}} = 1000\ mL\ or\ 1\ liter$$

PROBLEM SET 2

CHARLE'S LAW

As previously stated, Charle's Law deals with temperature. An increase in temperature causes an increase in volume, pressure remaining constant.

Note: When gas cools its volume decreases by $\frac{1}{273}$ of its original volume at 0°C for every 1°C drop in temperature. The same is true for an increase in temperature. One volume ↕ by $\frac{1}{273}$ at 0°C for every 1°C.

Example 3

What volume will 300 mL of H_2 occupy if temperature is increased from 40°C to 60°C?

Solution:

1. Convert temperature to Kelvin: °C + 273K   40 + 273 = 313 K   60 + 273 = 333 K
2. $V_2 = \frac{V_1 \, x \, T_2}{T_1} = \frac{300\ ml\ x\ \cancel{333\ K}}{\cancel{313\ K}} = 319.2\ mL$

Example 4

(CHARLE'S LAW)

A chemist has 625 mL of O_2. He decreases temperature from 30°C to 5°C. Find the volume.

Solution: $V_2 = \frac{V_1 \, x \, T_2}{T_1} = \frac{625\ ml\ x\ 278\ \cancel{K}}{303\ \cancel{K}} = 573.4\ mL$ *(Note: K or A constants 273 = °C)*

COMBINED GAS LAWS (BOYLE'S AND CHARLES')

The combined gas laws are important because it is difficult to maintain constant temperature and pressure in the lab.

EXAMPLE 5

What volume will 950 mL of H_2 gas occupy if the temperature is increase from 40°C to 60°C and pressure is decreased from 800 torr to 760 torr?

Solution: $$V_2 = \frac{V_1 \, x \, T_2 \, x \, P_2}{T_1 \, x \, P_1} = \frac{950 \, ml \, x \, \cancel{333 \, K} \, x \, \cancel{800 \, torr}}{\cancel{313 \, K} \, x \, \cancel{760 \, torr}} = 1064 \, mL$$

EXAMPLE 6

A chemist has 600 mL of chlorine gas. She records a temperature of twenty five degrees Celsius and a pressure of 741 torr. She changes the pressure and temperature to STP. Find the new volume.

Solution: STP = 0°C or 273 K, 760 torr

$$V_2 = \frac{V_1 \, x \, P_1 \, x \, T_2}{P_2 \, x \, T_1} = \frac{600 \, ml \, x \, \cancel{741 \, torr} \, x \, \cancel{273 \, K}}{\cancel{760 \, torr} \, x \, \cancel{298 \, K}} = 536 \, mL$$

DALTON'S LAW AND COMBINED GAS LAWS

Dalton states that when two or more gases mix but do not combine the total pressure is equal to the sum of each gas involved. For example, air is 21% oxygen and 79% nitrogen. The partial pressure is 21 parts O_2 to 79 part N_2.

Simplified

When collecting O_2 in the lab, a student displaces water. Obviously, water and oxygen mix but do not combine. The result is oxygen and water vapor. Dalton's Law enables us to dry a gas using arithmetic.

EXAMPLE 7

400 mL or 400 O_2 are collected by the displacement of water at a temperature of 23°C and pressure of 795 torr. Find the volume of the dry gas at standard conditions.

refer to solution A

MOLAR VOLUMES AND WEIGHTS OF GASES

22.4 liters = 1 mole of gas
How many moles in 5 liter of H_2 gas?
Solution: we divide by the molar constant which is 22.4 liters = 1 mole of gas

$$5 \, \cancel{liter} \, H_2 \, x \, \frac{1 \, mole}{22.4 \, \cancel{L}} = 0.223 \, moles$$

GRAHAM'S LAWS

Graham is famous for describing the diffusion of gases. By diffusion we mean molecules from greater to lesser concentration. For example, an ammonia diffuses rapidly. When one cleans a surface, the ammonia diffuses throughout the area. Let's be more specific.

Diffusion of gases id proportional to weights. The lighter the gas, the faster it diffuses. On the other hand, heavier gases move slower.

Example Which gas will diffuse the fastest and slowest?

H_2 He Cl_2 O_2 F_2

H_2 is the fastest, lightest = 2 grams, **Cl_2** is slowest, heaviest =71 grams

RATES OF DIFFUSION – GRAHAMS'S LAW

Example

What are the relative rated of H_2 and Cl_2? The solution is simple $\frac{mass\ of\ 2}{mass\ of\ 1}$ and then find square root.

Solution $\frac{Cl_2}{H_2} = \sqrt{\frac{M\ W\ of\ 2}{M\ w\ of\ 1}} = \frac{71\ g}{2\ g} = 35.5\sqrt{35.5} = 5.95$ so H_2 moves 5.95 : 1 faster than Cl_2

DENSITY OF GASES

Density = mass/volume of D = M/V
What is the density of N_2 at STP?

Solution: Mass of N_2 = 28 grams and volume at STP is 22.4 liters
Density = M/V

$$\frac{28\ g\ N_2}{1\ mole} x \frac{1\ mole\ of\ a\ gas}{22.4\ liters} = 1.28\ grams/liter$$

Solution: (A)

Since water is involved, we must refer to the vapor pressure chart (see Figure 15-7 below). We find 23°C and notice the pressure in torr is 21.1 mm or torr. Now we subtract 795 torr – 21.1 = 773.9 torr.

Work as other combined gas law: $400\ mL\ x\ \frac{773.9\ torr}{760\ torr}\ x\ \frac{273\ K}{296\ K} = 375.6\ mL$

Figure 15-7

Vapor Pressure of H_2O at Various Temperatures

°C	torr (mm Hg)	°C	torr (mm Hg)
0	4.6	26	25.2
5	6.5	27	26.7
10	9.2	28	28.3
15	12.8	29	30.0
16	13.6	30	31.8
17	14.5	40	55.3
18	15.5	50	92.5
19	16.5	60	149.4
20	17.5	70	233.7
21	18.7	80	355.1
22	19.8	90	525.8
23	21.1	100	760.0
24	22.4	105	906.1
25	23.8	110	1074.6

DEVIATION FROM IDEAL GAS LAWS

Real gases do not conform to the gas laws. They only behave approximately to the laws because molecules have weight and occupy space. We also note that gases and molecule will attract each other, especially under high pressure and low temperature.

EXERCISE 15-1

1. State Boyle's Law and Charles' Law.
2. 300 mL of a gas at 760 torr pressure are compressed to 935 torr. At constant temperature, find the new volume.
3. 525 mL of a gas at 37°C are heated to 135°C at constant. Find the new volume of gas.
4. 225 mL of a gas will occupy what volume if temperature is heated from 30°C to 50°C and pressure is decreased from 800 torr to STP?
5. 768 mL of gas are collected over water at 40°C and 790 torr. Find its volume at standard conditions when dry (standard conditions are the same as STP constants).
6. What are the relative rated of diffusion of ammonia NH_3 and butane C_4H_{10}?
7. Calculate the relative rates of diffusion of H_2 and O_2.
8. How many molecules in 9 grams of sulfur dioxide (rotten egg odor) SO_2?
9. 2.45 grams of gas occupy 1.250 mL at STP. What is the weight of the gas?
10. Calculate the density of O_2, N_2, F_2, at STP.
11. What is the volume of 650 mL of H_2 if pressure changes from 960 torr to STP and temperature changes from 30°C to 50°C?
12. What volume will 425 mL of N_2O occupy if temperature is decreased from 60°C to 35°C pressure remaining constant?
13. What is the density of H_2S at STP (note: D = M/V)
14. What are the relative rates of diffusion of N_2 and Cl_2?
15. 7 liters of gas "x" weighs 12 grams. Find mass of gas x.

EXERCISE 15-2

A. A surgeon can increase the volume of anesthesia by __________ , pressure.
B. Patient ***X*** received 5.6 L of anesthesia or __________ moles or __________ molecules.
C. A technician increases the temperature of a gas from 15°C to 40°C. his volume will __________.
D. Letter A is __________ law and letter C is __________ law.
E. Which gas will diffuse the slowest __________ , and the fastest __________ ? CO_2, NO, H_2, SO_2, F_2, N_2
F. 9.046×10^{23} molecules of NH_3 is __________ moles, and __________ grams.
G. 180 liters of O_2 __________ moles.
H. 5.6 g of N_2 __________ moles, __________ molecules.

FOR FURTHER STUDY TOPICS

GAY – LUSSAC'S LAW

The pressure of a fixed mass of gas at constant volume is directly proportional to the Kelvin temperature:

P = kT (at constant volume) $\frac{P_1}{T_1} = \frac{P_2}{T_2}$

Simplified if ***temperature decreases → pressure will decrease***

Furthermore combining volumes of reacting gases and their products of gases can be related by small whole number ratios. $CH_{4(2)} + 2O_{2(g)} + 2H_2O_{(g)} + CO_{2\,(g)}$

1 2 2 ratio

A Sample Problem using temperature and pressure.

A container of H_2 has a pressure of 820 torr at 33°C. The container is sealed and cooled to 0°. Find the new pressure.

$$P_2 = P_1\, x\, \frac{T_2}{T_1} = 820\ torr\ x\ \frac{\cancel{273\ K}}{\cancel{306\ K}} = 731.6\ torr$$

Result: ***decrease pressure → decrease volume,***
decrease temperature → decrease pressure

AMADE – AVOGADO HYPOTHESIS

Equal volumes of gases under identical conditions of pressure and temperature contain the same number of molecules.

CHAPTER 16

EQUATIONS AND CALCULATIONS

Goals and Objectives

To be able to

- interpret and analyze chemical equations
- find volume and mass from reactants and products
- use coefficients as a major tool in equation solving

Key terms

- reactants
- products
- mass volume molar volume limiting reactant
- excess reactant
- moles
- yield

CHAPTER 16

EQUATIONS AND CALCULATIONS

Chemists can calculate and predict results from equations.

Example: mass to mass

How many grams of KCl will be produced by heating 36 grams of $KClO_3$?

Step 1: study balanced equation

$2KClO_3 \xrightarrow{heat} 2KCl + 3\ O_2$

Step 2: calculate equation weights and write given grams in question as numerator

$2KClO_3$ = 244 grams (denominator) and 36 grams are in questions or the numerator

$$36\ g\ KClO_3\ (s)\ x\ \frac{\cancel{1\ mole}\ KClO_3}{122.5\ g\ x\ \cancel{2\ moles}}\ x\ \frac{2\ x\ 74.5\ g\ KCl\ (s)}{\cancel{1\ mole}} = 21.8\ g\ of\ KCl\ (s)$$

How many grams of CuO will be produced from the decomposition of 20 grams $CuCO_3$?

$$CuCO_3 \xrightarrow{heated} CuCO + CO_2$$

Solution: we note equation is balanced

NOTE: Since 20 grams of $CuCO_3$ is in the question it becomes part of the numerator we should note that CUO (mass) is also a numerator because we need its mass change

Solution: $\frac{20\ g\ x\ \cancel{1\ mole}\ CuCO_3}{123.54\ g\ CuCO_3}\ x\ \frac{79.5\ g\ of\ CuO}{\cancel{1\ mole}} = 12.9\ g\ of\ CuCO_3$

VOLUME TO VOLUME EQUATIONS

$$N_2 + 3\ H_{2(g)} \rightarrow 2\ NH_{3(g)}$$

How many liters of NH_3 (ammonia) will 5 liters of N_2 produce?

Solution: Convert moles to liters. One mole of a gas 22.4 liters.

$$5\ liters\ N_2\ x\ \frac{\cancel{1\ mole}}{22.4\ liters}\ x\ \frac{\cancel{2\ moles}\ NH_3\ x\ 22.4\ L}{\cancel{1\ mole}} = 10\ liters\ of\ NH_3\ gas$$

VOLUME PROBLEMS

How many liters of O_2 will 27 grams of $KClO_3$ yield?

$$2\ KClO_3 \xrightarrow{heated} 2KCl + 3O_2$$

One mole of $KClO_3$ = 122.5 grams but equation has 2 mole. Therefore, 2 moles x 122.5 grams = 245 grams and 3 moles of O_2 = (22.4 liters x 3 moles) or 67.2 liters. Note: factoring in solution.

Solution: $\frac{27\ \cancel{g}\ of\ KClO_3\ x\ 1\ \cancel{mole}}{2\ \cancel{moles}\ x\ \cancel{122.5g\ KClO_3}} \times \frac{22.4\ L\ of\ O_2\ x\ \cancel{3\ moles}}{\cancel{1\ mole\ of\ O_2\ (g)}} = 7.4\ liters$

More Examples

A chemist electrolytes KCl and records 76 liters of Cl_2. How many grams of KCl did he use?

$$2KCl \rightarrow 2K + Cl_2\ (g)$$

$$\frac{\cancel{76\ L}\ of\ Cl_2\ x\ \cancel{1\ mole}}{\cancel{22.4\ L}} x \frac{74.5\ g\ x\ \cancel{2\ moles}}{\cancel{1\ mole\ KCl}} = 505.53\ grams\ of\ KCl$$

MOLES TO MOLES

$$2NaNO_3(s) \rightarrow 2NaNO_2(s) + O_2(g)$$

Four moles of $NaNO_3(s)$ yield ________ moles of $NaNO_2(s)$

Solution: $\frac{4\ moles\ NaNO_3}{2\ moles}\ x\ 2\ moles\ NaNO_2 = 4\ moles\ of\ NaNO_2$

For Further Study

The ***limited reactant*** is important in equation calculations because its amount sets a limit on the yield of product. An example of an equation used to make fertilizers for lawns and gardens.

$$N_2\ (g) + 3H_2 \_\_\_\_\_\_\_\_ 2NH_3\ (g)$$

How much ammonia is produced if 12 grams of N_2 reacts with 8 grams of hydrogen.
First find the ***limiting reactant***, if we convert grams to moles.

$$12\ \cancel{grams\ of\ N_2} x \frac{1\ mole\ N_2}{28\ \cancel{g\ of\ N_2}} = 0.4285\ moles$$

$$8\ \cancel{grams}\ of\ H_2\ x \frac{1\ mole\ H_2}{2\ \cancel{g}\ x\ 3\ moles} = 1.33\ moles$$

N_2 is limited reactant because 0.4285 moles is less than H_2

Final Solution:

How much ammonia is produced?

N_2 (g) + $3H_2$ (g) ———— 2 NH_3 (g)

$$\frac{8\ \cancel{grams}\ of\ \cancel{1\ mole\ N_2}}{28\ \cancel{g}}\ x\ \frac{2\ \cancel{moles}\ NH_3\ x\ 17\ g}{\cancel{1\ mole}} = 14.57\ grams\ of\ NH_3$$

$$Volume\ of\ NH_3 =\ 14.57\ \cancel{g}\ x\ \frac{\cancel{1\ mole\ NH_3}\ x\ 22.4\ L}{17\ \cancel{g}\ x\ \cancel{1\ mole\ of\ gas}} = \mathbf{19.2\ liters\ of\ NH_3}$$

Exercise 16-1

1. How many grams of Fe_2O_3 will be performed by the action of 1.5 moles of Fe?

$$4\ Fe + 3\ O_2 \rightarrow 2\ Fe_2O_3$$

2. How many grams of Fe_2O_3will be formed by the action of 60 grams of O_2? (see above equation)
3. How many moles of Fe_2O_3will be formed by 73 grams of O_2? (see above equation)
4. A chemist burns 6 grams of $KClO_3$. How many liters of O2 are produced?

$$2\ KClO_3 \rightarrow 2\ KCl + 3\ O_2$$

5. The equation below is for the burning of acetylene in air. How many grams of water will be formed from 3.5 moles of C_2H_2?

$$2\ C_2H_2 + 5\ O_2 \rightarrow CO_2 + 2\ H_2O$$

6. How many liters of CO_2 will be formed from 3.5 moles of C_2H_2 (acetylene)?
7. 0.6 moles of C_2H_2 will yield ________ liter of CO_2?
8. 60 grams of NaCl will yield ________ grams of Na_2SO_4

$$2\ NaCl + H_2SO_4 \rightarrow Na_2SO_4 + 2\ HCl$$

9. A chemist has 35.5 grams of Na_2SO_4. How many grams of H_2SO_4 did he use?
10. The limited reactant in equation 8 is ________, the excess reactant is ________.

CHAPTER 17

EQUILIBRIUM

To be able to:

- explain the concepts of equilibrium
- study the effects of temperature, catalysts and nature of chemical reactants
- solve some equilibrium problems
- predict directional shifts
- analyze energy and equilibrium diagrams, tablets and charts

Key terms:

- activated complex
- activation energy
- catalyst
- endothermic
- exothermic
- ionization constants
- solubility product constants (Ksp)
- precipitate
- mole per liter

Chapter 17

EQUILIBRIUM

In case of a weak electrolyte, we have two reactions occurring simultaneously in opposite directions. First the dissociation takes place faster than the recombination, but eventually as the ions increase, the recombination builds up to equal the rate of dissociation. After both process proceed at the same rate in opposite directions, we have equilibrium.

The position of equilibrium depends upon the following:

A. Pressure
An increase in pressure would shift the equilibrium to the smaller volume.

$$N_2\,(g) + 3H_2 \leftrightarrows 2NH_3\,(g)$$

B. Temperature
An increase in temperature will shift equilibrium to the elements absorbing energy (endothermic).

C. Catalyst
The addition of a catalyst will speed up equilibrium by lowering activation energy. *However, the catalyst itself is unchanged.*

D. Le Chateliers Principle
If a stress is applied to a reaction at equilibrium to the equilibrium will be displaced in the direction that will relieve the stress.

E. Law of Mass Action
The Law states that the velocity of a reaction is proportional to the product of the molar concentrations of the reacting substances taken to the proper powers. The powers are the co-efficient of the reactants in the balanced chemical equation for reaction.

Simplified

The Law states that the greater the molarity, the more velocity of equilibrium.

Expression Keq

$$A + B \leftrightarrows C + D$$

$$\frac{[C]\,x\,[D]}{[A]\,v\,[B]} = \frac{Products}{Reactants}$$

Note: The left side is usually the denominator and the right side is the numerator. The brackets indicate moles per liter.

EQUILIBRIUM EXAMPLES

Write the equilibrium expression for the following equation.

Reactants $N_2(g) + 3H_2CH(g) \leftrightarrow 2NH_3\ (g)$ *Products*

$$\frac{[NH_3]^2}{[N_2g][H_2]^3}$$

Note: Reactants are numerators and products are denominators. The molar concentrations are raised to exponents.

Write the equilibrium for the following equation.

$$mC + nD = pE = qF$$

$$keq = \frac{(E)^p\ x\ (F)^q}{(C)^m\ x\ (D)^n}$$

We analyze the above reaction. The keq or rate of reaction to the left is

$$N_2 = K_2\ x\ (E)^p\ x\ (F)^q$$

N^1 is the rate of velocity and K^1 is the constant of proportionality.
At equilibrium the two rates will be equal.
If $K_2\ x\ (E)^p\ x\ (F)^q = K^1\ x\ (C)^m\ x\ (D)^n$

Then $\frac{(E)^p\ x\ (F)^q}{(C)^m\ x\ (D)^n} = \frac{K_1}{K_2} = K$

The constant K_1 is a constant for a given.

EXAMPLE 1

A closed container containing PCl5 was heated to 185° until equilibrium was established. Analysis showed the following.

$$PCl_5(g) \leftrightarrow PCl_3(g) + Cl_2(g)$$

PCl_3 = 0.086 moles per liter, Cl_2 = 0.086 moles per liter, PCl_5 =0.25 moles per liter. Calculate Keq.

Solution: Write a Keq expression.

$$K = \frac{(PCl_3)(Cl_2)}{(PCl_5)} = \frac{(0.086)(0.086)}{(0.25)} = 0.0296\ at\ 185°C$$

EXAMPLE 2

$$Cl_5(g) \rightleftarrows PCl_3(g) + Cl_2(g)$$

A chemist placed quantities of PCl_3 and Cl_2 in a chamber and heated it to 185°C at one atmosphere until equilibrium was reached.

PCl_3 = 0.265 moles per liter, Cl_2 = 0.204 moles per liter, Calculate concentration of PCl_3.

Solution: K = 0.0296 (from previous problem)

$$PCl_3 = \frac{K\ x\ (PCl_5)}{Cl_2} = \frac{0.0296\ x\ \left(\frac{0.265m}{1}\right)}{0.204} = 0.0384\ moles\ per\ l$$

EXAMPLE 3

A liter of 2HI was heated until equilibrium was established. 2HI ↔ H_2 + I_2. Analysis showed the following molarity:

(H_2) = 0.36 moles per liter, (I_2) = 0.36 moles per liter and (HI) = 2.16 moles per liter. Calculate K for the above equation.

$$K = \frac{(H_2)(I_1)}{HI^2} = \frac{(0.36)(0.36)}{(2.16)^2} = 0.0278$$

MORE USES OF EQUILIBRIUM

The ionization constants are defined as the concentration of both ions of weak electrolytes and undissociated molecules. Ionization constants are usually small number such as 10^{-2} or as small as 10^{-12}. The K_a or ionization constant is a measure of the extent to which a weak acid ionizes in water.

EXAMPLE 1

At 25°C acetic acid $HC_2H_3O_2$ is 1.34% ionized in 0.1M solution. What is its ionization constant?

$$HC_2H_3O_2 - H^+ + C_2H_3O_2$$

$$\frac{(H)(C_2H_3O_2)}{(HC_2H_3O_2)}$$

A. Convert 1.34% to decimal = 0.0134 x molarity which is 0.1M
B. Subtract m-decimal = 0.1 – 0.00134 = 0.09866
C. Substitute number of moles in Keq expression $\frac{(0.00134)(0.00134)}{(0.09866)} = 1.82\ x\ 10^{-5}$

EXAMPLE 2

A 0.5M H_3PO_4 solution is 16% ionized according to the equation $H_3PO_4 \rightarrow H^+\ H_2PO_4$ at 25°C. Find K_a.

Solution:

1. Write Keq $\frac{[H^+]\,[H_2PO_4]}{[H_3PO_4]}$
2. Convert % to decimal and multiply by molarity = 0.16 x 0.5M = 0.08
3. Subtract molarity from decimal 0.5M – 0.08 = 0.42

$$\frac{(0.08)^2}{0.42} = \frac{0.0064}{0.42} = 0.0152\ V\ 1.52\ x\ 10^{-2}$$

KSP

Most of us believe we can look at a substance in a solution and describe its degree of solubility. Alka Seltzer, soap, powered soft drinks and most medications are extremely soluble in water. However, we know that all substances have a certain degree of solubility. Aluminum containers do not mix with food or do they? Will water dissolve the liquid that holds diamonds in place in a ring setting? Chemists must know the solubility of substances to answer these questions.

A more positive exponent is extremely soluble. On the other hand, the smaller the negative exponent, the more insoluble. Ksp is define as solubility of saturated solutions of slightly soluble electrolytes. Assume 100% ionized in solution.

Example: 2.0×10^{8} soluble in H_2O HgS 1.0×10^{-52} insoluble in H_2O

KSP PROBLEMS

EXAMPLE 1

Write Ksp expression for AgCl Ag^{+} (aq) = Cl (aq)

Solution: [Ag] x $[Cl^{-}]$

EXAMPLE 2

If concentration of Ag is 1.3×10^{-5} moles per liter and Ag is equal to Cl^{-}. What is Ksp?

Solution: Ksp expression is [Ag] x $[Cl^{-}] = (1.3 \times 10^{-5})^2 = 1.7 \times 10^{-10}$ Ksp of AgCl

EXAMPLE 3

The Ksp of CdS 10×10^{-27}. Find the concentrations of the ions of this salt.

Solution: $CdS \rightarrow Cd^{+} + S^{-} = (Cd^{+}) \times (S^{-})$

Since we have the Ksp and two ions we take the square roor of the Ksp which is 1.0×10^{-27}

$$\sqrt{1.0 \times 10^{-27}} = 3.16 \times 10^{-14} \text{ moles per } l$$

Proof: $(3.16 \times 10^{-14})^2 = 1.0 \times 10^{-27}$

EXAMPLE 4

The Ksp of $PbCl_2$ is 1.6×10^{-5}. Calculate the solubility in grams per liter.

Solution: This problem, only asks for g/l of $PbCl_2$. However, $PbCl_2$ has 3 ions. $PbCl_2 \rightarrow 2Cl + Pb$, therefore, we must divide by 4 and then get the cube root. Why?

Note: $PbCl_2 = Pb^{++} + 2Cl^{-} = Pb = 1.6 \times 10^{-5}$

$2Cl = 1.6 \times 10^{-5} \times 2 = 3.2 \times 10^{-5}$, but moles are exponents

$= (3.2 \times 10^{-5})^2$ or x $(2X)^2 = 4x^3$

Since we must do an inverse, we have the following for $PbCl_2$.

$$\frac{1.6 \times 10^{-5}}{4} = 4.0 \times 10^{-6}$$

$$3\sqrt{4.0 \times 10^{-6}} = 1.59 \times 10^{-2} \text{ molar solubility}$$

Our final answer should be in grams per liter. Therefore, molar solubility x wt of $PbCl_2$ = $1.59 \times 10^{-2} \times 278$ (1 mole) = **4.39 grams per liter.**

EXAMPLE 5

A chemist has 2.89×10^{-6} grams per liter of AgCl. Hu must find the Ksp.

Solution: AgI → Ag x I

Step 1: Convert *grams per liter* to *mole per liter* to solve for Ksp.

Wt. of AgI is 235 grams

$$\frac{Grams}{moles} = \frac{2.89\ x\ 10^{-6}}{235} = 1.23\ x\ 10^{-8}$$

Since AgI → [Ag][I] we get $(1.23 \times 10^{-8})^2 = 1.5 \times 10^{-16}$

Ksp of AgI is 1.5×10^{-16}

EXAMPLE 6

Find the Ksp of PbI_2 if an engineer has 7.0×10^{-1} grams per liter. PbI_2 → Pb + 2I

Solution:

Step 1: Convert to moles per liter

$$\frac{7.0\ x\ 10^{-1}}{461\ wt\ pf\ PbI_2} = 1.52\ x\ 10^{-3}\ moler\ per\ l$$

Step 2: [Pb] 1.52×10^{-3}[2I] $(3.04 \times 10^{-3})^2 = 1.52 \times 10^{-3} \times 9.24 \times 10^{-6}$

= **Ksp of 1.4×10^{-8} for PbI_2**

PRECIPITATES

Chemists use the term precipitate to describe an insoluble substance or two liquids resulting in a solid or gel. In order to get a precipitate the Ksp must exceed the original substance Ksp.

EXAMPLE 7

A solution contains both chloride ions Cl^- and the chromate ion CrO_4^-. The (Cl^-) is 0.02M and the (CrO_4^-) is 0.08M. a chemist adds silver nitrate to the solution.

The Ksp of AgCl is 1.6×10^{-10}. The Ksp of Ag_2CrO_4 is 9.0×10^{-12}.

A. What molarity will each compound start precipitation?
B. Which one will precipitate first?

Solution: $AgCl = (Ag^+) \times (Cl^-) = X \times 0.02 = 1.6 \times 10^{-10} = \frac{1.6 \times 10^{-10}}{0.02}$ = **8.0 x 10^{-9} mole per liter** (precipitation occurs)

$Ag_2CrO_4 = (Ag)^2 \times (CrO_4^-) = 9.0 \times 10\text{-}12 = \frac{9.0 \times 10^{-12}}{0.08}$

$X^2 \times 0.08 = 9.10 \times 10^{-12}\ 1.12 \times 10^{-10}$

$\sqrt{1.12 \times 10^{-10}} = \mathbf{1.06 \times 10^{-5}}$

The silver chloride AgCl will precipitate first because of its greater insolubility, 8.0 x 10^{-9}, whereas Ag_2CrO_4 is more soluble or 1.06 x 10^{-5}.

Exercise 17-1

1. Calculate Ka, ionization constant for HF, given that 7% ionized in M solution.
2. Write Keq expression for $2S_{(g)} + 20_{(g)} \rightarrow 2SO_{2(g)}$
3. Solve for Ksp: 1.65 x 10^{-4} g/l of AgBr, $BaSO_4$ 1.0 x 10^{-5} moles per liter
4. Give solubility in molar concentrations and grams per liter.
   $PbCrO_4$ Ksp = 1.8 x 10^{-14} AgCl Ksp = 1.6 x 10^{-10}

CHAPTER 18

ACIDS AND BASES

To be able to:

- study Bronsted and Lowry acids and bases
- use a logarithm table
- solve for pH and pOH
- study some common acids and bases
- master pH vocabular
- apply pH concepts to some chemical applications

Key words:

- acids
- amphiprotic
- bases
- buffer
- hydronium ion
- hydroxide ion
- molarity
- neutralization reaction
- percent of ionization
-

CHAPTER 18

ACIDS AND BASES

Acids and bases are part of our daily lives. Some common household bases are baking soda, detergents, soaps, toothpaste, and ammonia. Other bases are baking powder and Drano or sodium hydroxide (NaOH).

What is an acid? Acids are defined as proton donors (+) with hydrogen donating a positive charge in aqueous solutions that bonds with water forming a hydronium ion (according to Bronsted and Lowry). Chemists indicate some acids with the hydronium ion or H^+. the hydronium ion is also written as H_3O^+. acids also taste sour and turn litmus paper red.

Formation of a hydronium ion $H:\ddot{O}:H + H^+ \longrightarrow [H \; \ddot{O} : \; H]^+$ (H H)

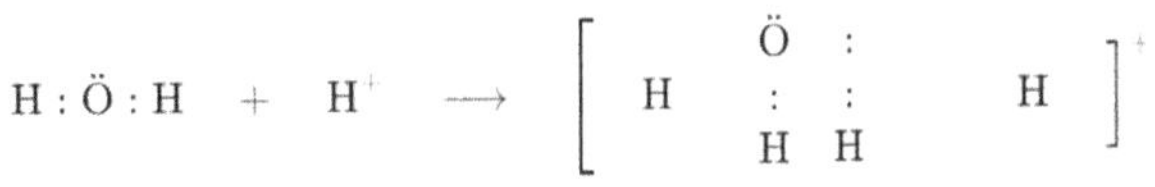

Formation of a hydronium ion, H_3O^+

What is a base? Many bases are slippery and taste bitter. Bronsted and Lowry define bases as proton acceptors. Several bases have the OH^- ion (hydroxide ion) present. Some bases such as ammonia NH_3 and baking soda $NaHCO_3$ do not have the hydroxide ion. Bases turn litmus paper blue.

An example of bases as proton acceptors. NH3 has an unshared pair of electron and can accept a proton (H^+) to form NH_4^+.

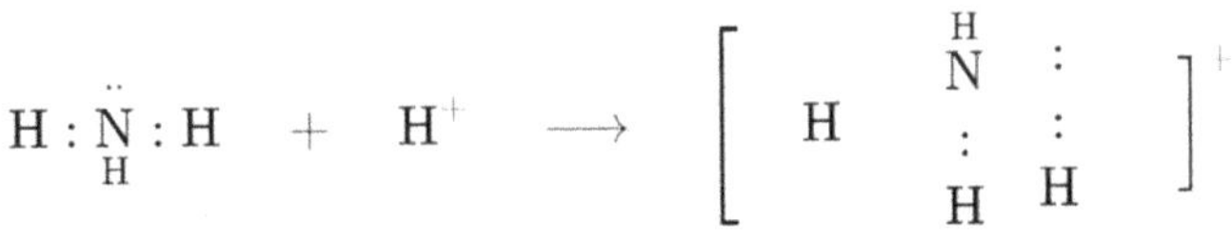

$$H:\ddot{N}:H + H^+ \longrightarrow [H \; N : \; H]^+$$

Amphoteric or amphiprotic

Compounds with acid and base traits are amphoteric or amphiprotic substances. Water is **amphoteric.** Water donates protons in lemon juice or acids and accepts protons in soap solutions. **Lemon juice is an acid or the H^+ and the soap solutions is the OH^- or the base part of water. Simplified water is acidic in lemon juice and basic in soap solutions.**

Another example of an amphiprotic substance is HCO_3^-. It acts as an acid and will lose a proton in a basic solution, according to the reaction: $HCO_3^- + OH^- \rightleftarrows CO_3^{2-} + H_2O$. In an acidic solution, however, HCO_3^- acts as a base and will accept a proton, according to the reaction: $HCO_3^- + H^- \rightleftarrows H_2O + CO_2$.

pH refers to **acids** and **bases** and is defined as the negative logarithm of the hydrogen ion concentration.

$pH = -(\log [H]^+)$ example if H^+ is 1.0×10^{-5} mole/liter, the pH is 5 or (5- log of 1.0)

$[H]^+ = 10^{-(pH)}$ example if pH is 4 H^+ is 1.0×10^{-4} mole/liter

Note: Since the pH scale is logarithmic acidity difference is not constant. A solution with a pH of 2 is not twice as acidic as one with a pH of 3 it is ten times as acidic. A solution with a pH of 1 is 100 times more acidic than a solution with pH of 3.

Hydrogen Concentration

The higher the concentration the stronger the acid.

1.0×10^{-0} to 1.0×10^{-6}	acids
1.0×10^{-7}	neutral
$1.0 \times 10^{-8} \rightarrow 1.0 \times 10^{-14}$	bases

Binary Acids – contain two elements
Ternary Acids – contain three elements
Other acids – Binary acids (containing two elements) can be made by the direct combination of the element with hydrogen gas, such as this reaction for the formation of hydrobromic acid.

$$H_2(g) + Br_2(l) \rightarrow 2HBr(g)$$
$$HBr(g) + H_2O(l) \rightarrow H_3O^+(aq) + Br^-(aq)$$

Other acids like phosphoric acid are formed when a nonmetallic oxide and water react.

$$P_4O_{10}(s) + 6H_2O(l) \rightarrow 4H_3PO_4(aq)$$

Carbonic acid is formed in the same way, using carbon dioxide gas that is bubbled through water.

$$CO_2(g) + H_2O(l) \rightarrow H_2CO_3(aq)$$

It is the presence of CO_2 in the air that causes natural rainwater to be slightly acidic. Carbonated water for soft drinks is made by this same reaction. The CO_2 is kept in solution by increasing the pressure on the solution.

Acids which contain three elements are ternary acids. Table 20-2 shows the names and formulas of some common binary and ternary acids.

FORMULAS AND NAMES OF SOME COMMON ACIDS			
BINARY ACIDS		TERNARY ACIDS	
HCl	hydrochloric	H_2SO_3	sulfurous
HBr	hydrobromic	H_2SO_4	sulfuric
HI	hydroiodic	HClO	hypochlorous
HF	hydrofluoric	$HClO_2$	chlorous

H_2S	hydrosulfuric	$HClO_3$	chloric
		$HClO_4$	perchloric
		HNO_2	nitrous
		HNO_3	nitric

EXPRESSIONS

Acids and bases are usually expressed in one of the two ways. (1) as pH quatity defined as the negative logarithm of the hydrogen ion or pH = log (H) 2 as the concentration of hydrogen ions in moles per liter (H^+). Moreover, pH is generally dependent on the concentration of the acid or base and degree of ionization.

$$(H) = (HX) \times di$$

Note **H^+ = concentration HX = acid d.i. = degree of ionization** and **H^+ = concentration**

We must account for bases or the hydroxide (OH) and pOH of a solution. They are related to (H^+) and pH as follows.

Acids	0 — 6.9
Neutral	7 — 7.1
Bases	7.2 — 14

Water Constants

$H_2O \rightarrow H^+ = OH^-$

When water dissociated, we get the above equation. Note the constants.

$H^+ = 1.0 \times 10^{-7}$ OH $= 1.0 \times 10^{-7}$
Ki of $H_2O = (H^+) \times (OH) = 10^{-14}$

Strong Acids	Complete dissociation of H^+ examples: **$HClO_4$, HCl, HNO_2**
Strong Bases	Complete dissociation of OH examples: **NaOH, KOH, RbOH, LiOH**
Weak Acids	Partial dissociation of OH^- examples: **ammonia, baking soda, methylamine, aniline**
Conjugated Acids:	Occurs when proton is transferred to base.
Conjugated Base:	Occurs when an H+ is lost.

Conjugated Acids And Base Pair:

Occurs by donating and accepting a single proton, usually *hydrogen*.

Conjugated Acid	Conjugated Base
HCl	Cl^-
H_2SO_4	HSO_4

Note we usually reduce one hydrogen for conjugated base.
A strong acid has a weak conjugated base. A weak acid has a strong conjugated base.

Some Common Bases

Name	**Formula**	**Common Name**
Sodium hydroxide	NaOH	Lye or caustic soda
Potassium hydroxide	KOH	Lye or caustic potash
Magnesium hydroxide	$Mg(OH)_2$	Milk of magnesia
Calcium hydroxide	$Ca(OH)_2$	Slaked lime
Ammonia water	NH_4OH	Household ammonia

Some Common Acids

Formula	**Acids**	**Common Name**
HNO_3	Nitric	Aqua fortis
H_2SO_4	Sulfuric	Oil of Vitrol
$HC_2H_3O_2$	Acetic	Vinegar
H_2CO_3	Carbonic	Soda, carbonated water, pop
HCl	Hydrochloric	Muriatic acid

ACID AND BASE PROBLEMS

EXERCISE 18-1

1. What is the pH of a 0.1M solution of HNO_3 (nitric acid).
   Solution: Convert to H^+ or hydrogen ion concentration (scientific notation).
   $0.1M = 1.0 \text{ x } 10^{-1}$ $\quad$ $pH = -\log (H^+)$
   pH = 1 -log of 1.0 pH = 1
2. What is the pH of a 0.025M solution of HNO_3?.
   H^+ - 2.5 x 10-2 pH = 2 -log 2.5 pH – 1.6 (100% ionization)
3. The pH of a solution is 3.64. What is the H^+ or hydronium ion concentration?
   Step 1: Note **3.64** is a negative number, so we must raise it to a whole number or 4 (written as 10^{-4}).
   Step 2: Since 0.36 was added to 3.64 we take the inverse log of 0.36 which is 2.31 the H^+ = 2.3 x 10^{-4}
4. What is the percent ionization of a 0.1M acid solution whose pH is 4.3?
   Solution: $4.3 + 0.7 = 10^{-5}$ (0.7 inverse log is 5 so $H^+ = 5.0 \text{ x } 10^{-5}$)
   $$\% = \frac{5.0\,x\,10^{-5}}{0.01M}\,x\,100 = 0.5\%$$
5. The pH of a solution is 2. Find pOH.
   Solution: pOH refers to basic or hydroxide ion (OH) or alkali
   $$\text{Maximum base } 14 \text{ solution } \frac{1.0\,x\,10^{-14}}{1.0\,x\,10^{-2}}$$
   (quick solution) = 12 pH = 2 **(14 – 2 = 12 pOH)**
6. The pH of NaOH is 10. Find pOH.
   $$\frac{1.0\,x\,10^{-14}}{1.0\,x\,10^{-10}} = pOH\ of\ 4\ A\ pH\ of\ 10$$
7. Find the pH and H^+ of a 0.001M solution of acetic acid (HAc) that is 4%.
   Solution: $1.0 \text{ x } 10^{-3} \text{ x } 0.04$ $\quad$ $4.0 \text{ x } 10^{-5}$ $\quad$ **pH= 5 – 4.0 log, pH = 4.4**

BUFFERED SOLUTIONS

Buffers resist a change in pH. This is vital in our blood which must resist in pH from acids and base.

Buffers are made by using a weak acid and its salt (HF and its salt naF0. A weak base and its salt can also use (NH_3 and NH_2Cl).

BUFFERED SOLUTIONS PROBLEM

A solution contains 0.20M HF and 0.20M NaF. What is the pH of this solution?

Given: Ka of HF = 7.2 x 10-4 pH = 3.13

$$\frac{NaF = 0.20}{HF = 0.20} = 1\,x\,7.2\,x\,10^{-4}$$

ACIDS AND BASE PROBLEMS

EXERCISE 18-2

(Find pH for the following)

1. 0.1M HCl
2. 0.0001M KOH
3. 0.0267M HNO_3
4. 0.01M NaOH
5. 0.048M H_2SO_4

EXERCISE 18-3

(Find H^+ and Ph of the following)

A. Vinegar 0.003M and 3% ionized.
B. Boric acid 0.01M and 2.3% ionized.
C. Find pH if a chemist dissolves 0.09 grams of HCl in 50 ml of water.
D. A chemist dissolve 0.6 grams of NaOH in 80 ml of H_2O. What is the pH and pOH?
E. Buffered problem: A solution contains 0.60 M acetic acid and 0.58 sodium acetate. Ka of acetic: 1.8 x 10^{-5}. Find pH.
F. List some household acids and bases.
G. Shampoo Jacqui has a neutral pH. Write the H^+. what is the pH and pOH.

Find H^+ and pH of the following

H. Orange juice is acidic. The pH is between 9-12, 0-6.9, 7.2-9, none of these
I. Define salt, conjugate base, buffers, conjugated acid, litmus paper.

CHAPTER 19

NUCLEAR CHEMISTRY

Goals and objectives

To be able to:

- compare chemical and nuclear reactions
- study basic nuclear terms
- solve half life problems
- balance nuclear equations
- study some effects of radiation on living things

Key words:

- nucleus
- nuclear reactions
- alpha particles
- beta particles
- chain reactions
- gamma radiation
- positron
- radiation

CHAPTER 19

NUCLEAR SYSTEM

Nuclear reactions differ from ordinary chemical reactions in several ways.

A. In nuclear reactions the number of protons in the nucleus usually change, this would satisfy the chemist because they change elements into new ones.
B. Isotopes are unstable. Carbon 12 nucleus is stable whereas Carbon 14 isotope nucleus decomposes spontaneously.
C. Energy changes in nuclear reactions which is exceeded by several powers of 10. When carbon 14 decomposes to form nitrogen 14, an electron about one million kilojoules of energy is released.

In this chapter we will take a brief look at a few nuclear reactions.

TERMS

Alpha radiation – helium atom has a charge of +2 and a mass number of 4.

$^{4}_{2}He$

When an alpha particle is omitted, the atomic number decreases by two and the atomic mass decreases by four.

Example: $^{238}_{92}U \longrightarrow\ ^{4}_{2}He +\ ^{234}_{90}Th$

Beta particle – (B-) has a small mass and *increases atomic number by one.*

Example: $^{234}_{90}Th \longrightarrow\ ^{234}_{91}Pa +\ ^{0}_{-1}e$

Note that beta particles have great penetrating power. They can pass through walls or glass, but they cause less damage to human tissue than alpha particles.

Half-life is the time required for one-half of a radioactive sample to decompose. The half-life of a radioactive nucleus is constant, independent of concentration amount or any other factor.

HALF-LIFE PROBLEM

The isotope $^{90}_{38}$Sr is radioactive. It decays a half-life of 29 years. Suppose we start with a sample of weighing 0.0600g after 87 years.

What fraction of the sample is left?

Solution:

Step 1: First convert 87 years to half-lives. We divide 87 years by 1 half life = 3 half lives to 29 years

Step 2: Convert half lives to a fraction or decimal use formula use formula

use formula $\frac{(1)^n}{2}$ note n = number of half lives

substituting the formula we have $\frac{(1)^3}{2} = \frac{1}{8}$ $or\ 0.125$

fraction is left after 87 years

How many grams of strontium 90 remain after 87 years if you started with 0.600 grams?

Solution: **fraction x mass**

0.125 x 0.600 grams = 0.075 grams of Sr

A. NUCLEAR REACTORS

<u>Characteristics</u>

A. Devices that can regulate chain reactions involving nuclear fission.
B. Devices that can sustain nuclear reactions.
C. The reactor generates a tremendous amount of heat and is H2O cooled or cooled with air, helium, molten sodium and sometimes air.
D. Reactors produce more fuel than they consume.

DIAGRAM A – NUCLEAR REACTOR

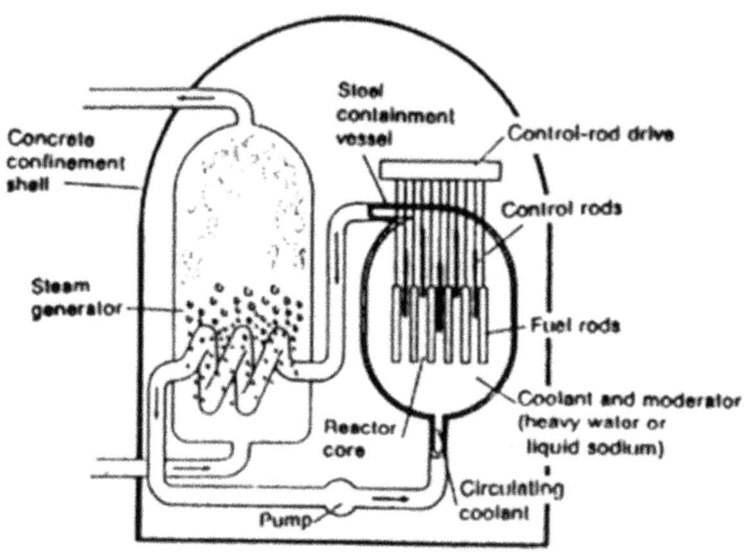

Schematic diagram of a nuclear reator.

B. CHAIN REACTIONS

1. Usually started by a neutron.
2. Product of one reaction triggers several reactions.
3. Maintained chain reactions increase energy.

DIAGRAM OF A CHAIN REACTION

Schematic diagram of the beginning of a chain reaction.

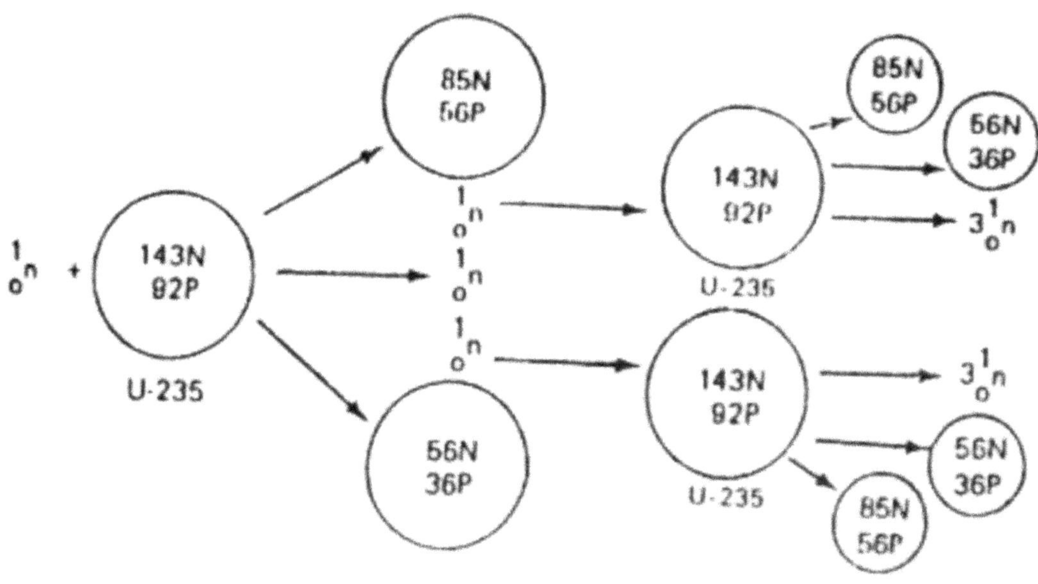

C. RADIATION – unstable

Nuclei decompose by themselves giving off energy.

1. Natural – elements found in nature that are unstable, example uranium.
2. Induced – unstable nuclei prepared in the lab. Examples: radium and polonium which were isolated from pitchblende by Marie and Pierre Currie.
3. Alpha radiation – positive charged particles.
4. Beta radiation – negative charge particles.
5. Gamma radiation – high energy, short wavelength and no change in atomic or mass number.

D. EFFECTS OF RADIATION

Harmful

1. Could induce cancer.
2. Birth defects
3. Change DNA
4. Exposure could be fatal

Beneficial uses

1. X-rays
2. Carbon dating
3. Cancer therapy

QUESTIONS

A. Describe a nuclear reactor. How are nuclear reactions controlled?
B. List some steps involved in a chain reaction.
C. What are the types of radiation? Describe some uses of radiation.

Nuclear Equations

Chemists use basic algebra to balance nuclear equations.

Basic rules:

- The sum of mass numbers are equal on both sides of the equation.
- The sum of electric charges sides are equal.
- Radiation particles given off during the equation are included in the equation.

Example A

238U→ 234Th + 4He
92 90 2

Note the mass numbers are equal on both sides and the electric charges are equal sample: A is an alpha particle.

Example B

Beta emission B – 0
-1e

234Pa → 234U + 0
91 92 -le

Note we have no change in mass numbers 234 and we have 91 charges on both sides because +92e -1e = 91e.

Example C

A positron B+
238U → 238Pa + 0
92 91 +1e

EXERCISE 19-2

Balance the following Nuclear Equations.

1. 238 → 234Th + ?
   92 90
2. 27Al + 4He → 30P + ?
   13 2 15
3. 226Ra → 4He + ?
   88 2
4. 14C → ? + 0e-
   6 -1

Note no change in mass numbers and algebraic rules total 92 electrons on both sides.

For Further Study: Purpose to find a half life

An 80 gram sample of radioactive element X decays to 20 grams in 36.8 seconds. What is the half life of this element?

The solution is simple. First convert grams to a fraction and then to half lives.

$$\frac{20\ grams}{80\ grams} = 0.25\ or\ {}^{1}/_{4}$$

To find half lives we need of 0.25 – 0.6 (-1) = 0.6. Since we need a half we need
Divide logs = 2 Since we started with **36.8 second we divide 36.8/2 = 18.4 sec half life**

For Further Study: Purpose to find mass in original nuclear sample

There are 73 grams of sample *x* left after 288 seconds. Find grams in the original sample. Half life is 57.6 seconds.

Solution:

Step 1 find half live 288 sec $x\ \frac{1\ half\ life}{57.6\ seconds} = 5\ half\ lives$

Step 2 half lives to decimal $(\frac{1}{2})^5 = 0.03125$

since we need the original sample mass we divide $\frac{given\ mass\ 73\ grams}{half\ life\ fraction\ 0.03125}$

original mass = 2336 grams

CHAPTER 20

Atomic Structure

Goals and Objectives

To be able to:

- study some atomic structures
- study orbitals
- draw and describe hybrid bonding
- study subshells study energy levels

Key words:

- protons neutrons
- electrons
- nucleus isotopes
- energy levels
- atomic number atomic mass
- hybrid bonding bond angles
- Lewis structures
- covalent bonds ionic bonds
- metallic bonds coordinate covalent bonding

CHAPTER 20

ATOMIC STRUCTURE

Earlier in our text we described how chemists use the periodic table. In this chapter we will give a brief explanation of bonding, shells, subshells, hybrid bonding and geometric orbitals.

TERMS

- Atoms – smallest particle of an element showing its properties
- two or more atoms of the same element or compound
- Nucleus – center of atom with protons and neutrons
- Shells – orbits in which electrons are formed
- Subshells – shells composed of orbitals
- Isotopes – same atomic number but different atomic weight
- Proton – atomic number, positive charge and equal to electrons (P)
- Ionic bad – opposite charges mainly in bonding salts and electrolytes
- Covalent – sharing of electron, mainly found in gases
- Coordinate covalent – one element donates most or all of the electrons or compound
- Lewis structures – dots that represents the outer most shell of an atom or compound
- Electronegativity – the attraction for electrons; the greater the difference in electronegativity, the more polar; the least differences, the more covalent (see electronegativity appendix)
- Neutrons – neutral charge (N)
- Electrons – negative charge (e-)
- Bohr atom – named after a pioneer scientist in atomic structure (Nelson Borh)
- C carbon (6P 6N)

  Note: 6 protons and 6 neutrons = atomic mass of 12
- O = oxygen (8P 8N)

  Oxygen has 8 protons and 8 neutrons = atomic mass of 26

SHELLS

There are seven shells: K, L, M, N, O, P and Q. The letters corresponds to approximate distance from the nucleus. The K shell is the first shell and its closest to the nucleus of the atom. The L shell, or second shell is one shell further from the nucleus. Thus, as the letters increase so does the distance from the nucleus increase.

Chemists can predict the maximum number of electrons in each shell by using L^2 x 2, where l = level or shell number.

Examples

How many electrons are in K shell?

K = first level l^2 x 2 = 2 electrons or 1^2, 1 x 2 = 2 electrons
L shell = 2^2 x 2 = 8 electrons
M shell = 3^2 x 2 = 18 electrons
N shell = 4^2 x 2 = 32 electrons
O shell = 5^2 x 2 = 50 electrons

The ***outermost shell*** of an element has the same number of electrons as its group number in the periodic table. Helium is an exception. It has two electrons in its outermost shell. It is in Group 8e or 18.

EXAMPLES OF BOHR ATOMS AND SHELLS

Li group 1	3P 4N	2e K	1e L

Atomic mass = 7, atomic number = 3 (also 3 electrons)
Note: Li is in group 1 and has one electron in the outer shell.

Ca group 2	20P 20N	2e K	8e L	8e M	2e N

Atomic mass 40, atomic number = 20
Note: Ca is in group and has two electrons in the outer shell.

Ca in group 2 has 2 electrons in its outermost shell. Carbon group 4 has 4 electrons, oxygen group 6 has 6 electrons. The outermost are in this shell. Chemical reactions can occur if the last shell has less than 8 electrons.

Shell and Structure (refer to periodic table)

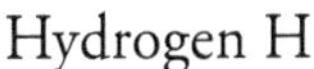

Hydrogen H

1e-
K shell

protons = electrons 1 proton and 1 electron

Carbon C

Note: K is full but L has room for 4 electrons

SUBSHELLS

Subshells are sub-levels within the levels. The subshells are important because they show that energy varies within each subshell level. The energy of electrons increase as the sub-level increase. Sub-levels are designated as s, p, d and f with s sub-level being lowest in energy and f being the highest. For comparison purposes 0.1 is the basement and f is the roof. Furthermore, sub-levels have a fixed capacity for electrons.

Sub-levels s, p, d, f

Electrons capacity s = 2, p = 6, d = 10, f = 14

SUB-LEVEL CONFIGURATION

7. Sp
6. Spd
5. Spdf
4. Spdf
3. Spd
2. Sp
1. S

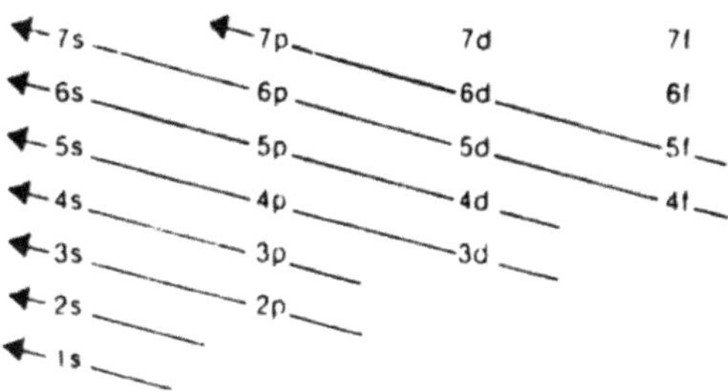

Note: The superscript indicates one electron in subshell and is the same as one electron in K shell. We should note the logical steps from 1-2-s3 levels, etc.

Note: The first subshell is under K, the second subshell has 4 electrons equal to L shell.

More Examples

Rectangles or orbital diagrams should have at least one electron (arrow) before two arrows are placed in rectangle.

Same Problem

Mg 12P 12N 2e 8e 2e

Group II

⇅	⇅	⇅	⇅	⇅	⇅
$1s^2$	$2s^2$	$2p^6$			$3s^2$

LEWIS STRUCTURE

Lewis Structures are dots that represent the last shell.

H The dot is for 1 electron in hydrogen's outer shell. H·

F the seven dots show Fluorine's outer shell of 7 electrons. :F̤̈:

C The carbon atom has 4 dots = 4 electrons in its outer shell. ·Ċ̣·

COVALENT BONDING, SHARING OF ELECTRONS

CH_4 or

```
      H
      |
H  —  C  —  H
      |
      H
```

The dash is for covalent bonds (2 electrons).

H2 = H:H   Gases are examples of covalent bonding

F2 = :F̤̈—F̤̈:

IONIC BOND

An ionic bond results when there is a complete transfer of electrons involving positive and negative charges. Ionic also exhibits a difference in polarity.

Ionic bond dissolves in water and conduct a current, however, it will not conduct as a solid. Salts are usually ionic bonds.

$$NaCl \rightarrow Na^+ + Cl^-$$

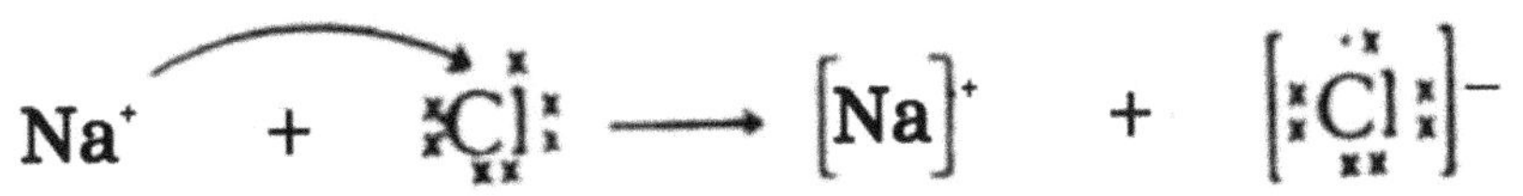

$$Na^{+} + \overset{\times}{\underset{\times\times}{\times Cl}}\text{:} \longrightarrow [Na]^{+} + [\text{:}\overset{\cdot\times}{\underset{\times\times}{Cl}}\text{:}]^{-}$$

Chlorine has 7 electrons and Na has one electron.

TRANSITIONAL ELEMENTS AND SHELLS

Transitional elements are found in group B and in group 8 of the periodic table. Electrons from the two outermost energy levels may be involved in a chemical reaction.

Moreover, most but not all transitional elements have 2 electrons in the outermost shell. Transitional elements have multiple oxidation states and are colored ions in solid and solution.

Example

Zn	30P	2e	8	18	2
	30N	K	L	M	N

Zinc atomic mass 65
Atomic number 30

Note: As the N shell fills with two electrons, a new transitional series starts.

COORDINATE COVALENT

$NH_3 +$ H–N̈–H (with H below N) $H_3 = \cdot\dot{N}\cdot$ ammonia gas molecule

Note: N has 5 electrons or 2 free electrons. Coordinate covalent shows us how one element contributes electrons and has some free. In coordinate covalent bonding one element does most of the work or supplies the majority of electrons.

BONDING ANGLES

Tetrahedral 109.5

Four pairs of electrons around central atom is tetrahedral of 109.5° (paired or unpaired)

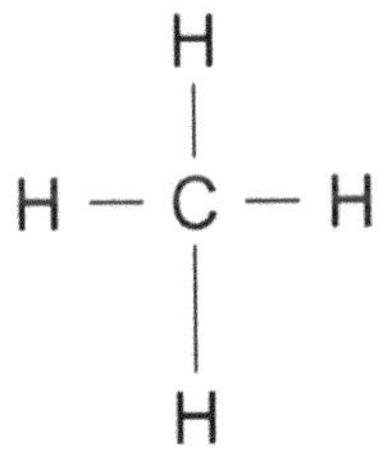

```
    H
    |
H — C — H
    |
    H
```

Planar – one or two pairs of electrons around central atom is linear 180°.

:F—Be—F: BeF_2

Trigonal – three pairs of electrons is trigonal or 120°

F—B—F :F—B—F:
| |
F :F:

or BF3

Figure 20-1

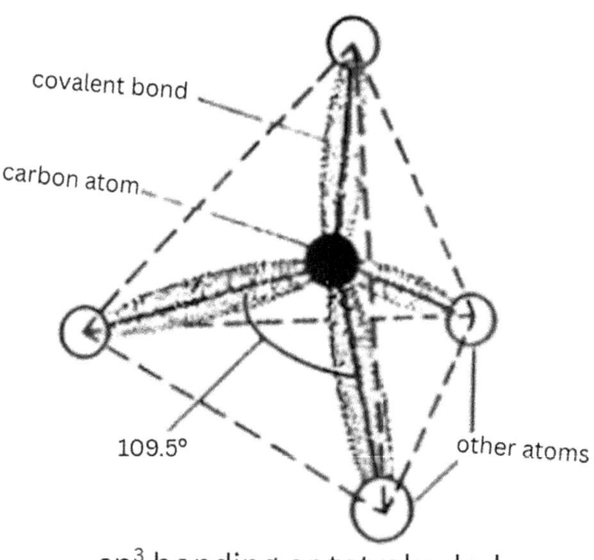

sp^3 bonding or tetrahedral

SHAPES

Chemists often predict how close electrons are to the nucleus. The 1s subshell is closer than 2s or 3s. the higher the energy level, the further the electrons are from the nucleus because the radius increases energy levels.

Figure 20-2

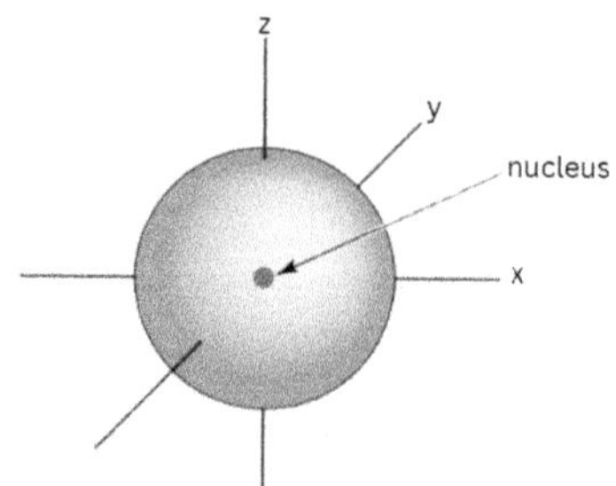

The three **p** orbitals or 6 elements are designed P_X P_Y P_Z. We know that each orbital holds one electron before a second electron is added.

XYZ SUBSHELLS

Example 1: Cl at number 17 $1s^2$
$2s^2$ $2px^2$ $2py^2$ $2pz^2$
$3s^2$ $3px^2$ $3py^2$ $3pz^1$

Since 3pz is unfulfilled, pz is available for bonding

Example 2: Nitrogen atomic number 7
$1s^2 2s^2$ $2px^1$ $2py^1$ $2pz^1$

Hybrid bonding. Carbon atomic number =6
$1s^2$ $2s^2$ wpx^1 $2py^1$

When carbon bonds with hydrogen, one of the 2s2 electrons is promoted to a $2pz^1$

Figure a-1

$1s^2$	$2s^2$	$2px^1$	$2py^1$
⇅	⇅	↑	↑

This diagram (Figure a-1) shows an empty orbital. A carbon 2s electron leaves the 2s orbital and enters the empty orbital when approached by hydrogen.

Figure a-2 (these four orbitals are called sp^3 orbitals)

$1s^2$	$2s^1$	$2px^1$	$2py^1$	$2pz^1$
⇅	↑	↑	↑	↑

S + 3p = sp^3 hybrid bonding

Another example of hybrid bonding is BF3, B is the central atom $1s^2$ $2s^2$ $2px^1$

$1s^2$	$2s^2$	$2px^1$
⇅	⇅	↑

Since $2py^1$ can take one or more electrons F cause the bonds to overlap, thus forming sp^2 hybrid:

$1s^2$	$2s^1$	$2px^1$	$2py^1$
⇅	↑	↑	↑

(Note sp2 hybrid bonding)

Diagram A

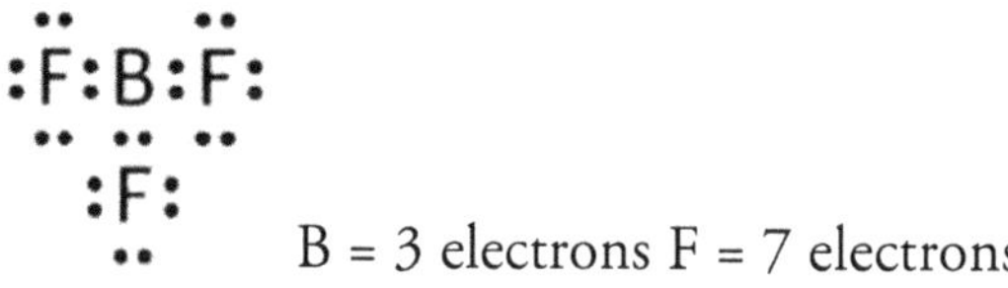

B = 3 electrons F = 7 electrons

Figure 20-3

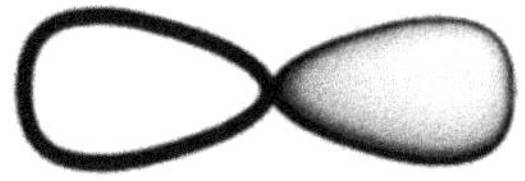

Figure 20-4

Fig. 7-7. Orientation in space of the p orbitals.

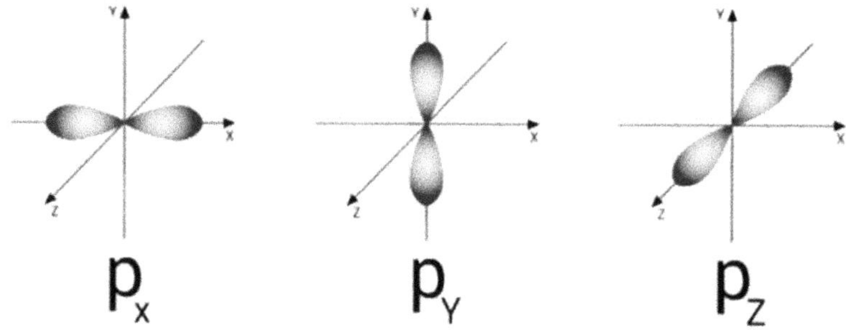

Figure 20-5

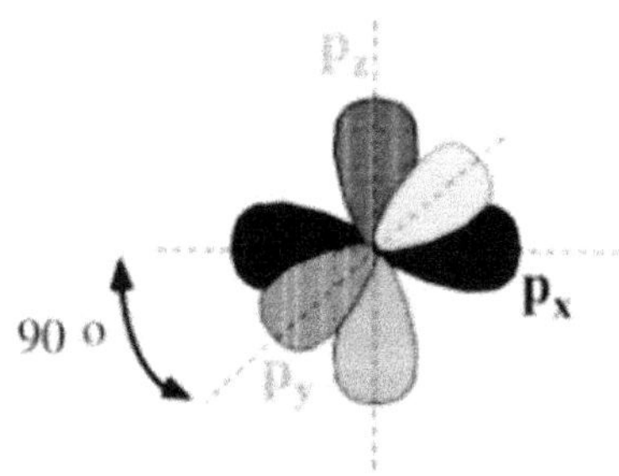

Figure 20-6
Some hybrid bond orbitals.

HYBRID BOND ORBITALS	GEOMETRIC SHAPE	HOW FORMED	SOME MOLECULES (CENTRAL ATOMS ARE BOLD FACED)
Linear *sp*	180°	From one *s* and one *p* electron	BeH_2
Trigonal planar sp^2	120° 120° 120°	From one *s* and two *p* electrons	BF_3, AlH_3
Tetrahedral sp^3	109.50	From one *s* and three *p* electrons	CH_4, $SiCl_4$, SO_4^{2-}, NH_4^+

EXERCISES 20-1

(Review Questions)

1. Draw atomic orbitals and subshells for oxygen, calcium, chlorine, copper +2, zinc, hydrogen, and magnesium.
2. Describe hybrid bonding in CH_4 and NH_3.
3. Define shells, subshells, atomic orbitals, isotopes, atoms, molecule, hybridization.
4. Describe bond angles in sp, sp^2, sp^3.
5. Draw Lewis structures for KBr.
6. Draw Lewis structures for H_2SO_4, H_2, F_2, KCl.
7. An element is $1s^2\ 2s^2\ 2p^6\ 3s^2$. What is the group number? Name the element.
8. Explain why CsF or cesium fluoride exhibits electronegativity.

CHAPTER 21

SPECIAL TOPICS

Goals and Objective

To be able to:

- apply general chemistry concepts
- write formulas
- solve for oxidation numbers
- name compounds
- energy word problems
- Nernst Equation

Key terms:

- formulas
- energy
- polyatomic ions
- Law of Hess
- Nernst equation

CHAPTER 21

SPECIAL TOPICS

WRITING FORMULAS

Writing charge or combining capacity over elements and cross.

Examples: H is in group 1 or +1. Oxygen is in group 6 or -2.

$H^{+1} + O^{-2} = H_2O$ (note crisscross) $Al^{+3}\ O^{-2} = Al_2O_3$

$Al^{+3}\ SO_4^{-3}$ $Al_2(SO_4)_3$ (Use parentheses for radicals)

$Fe^{+3}\ O^{-2} = Fe_2O_3Cu^{+2}Cl^{-1} = CuCl_2$

NAMING COMPOUNDS

A. The group I, II and III elements are often compounds with groups V, VI and VII. Chemists use the metal name and the non-metals end in ide.

Examples:

- NaCl – sodium chloride
- $CaCl_2$ – calcium chloride
- HF – hydrogen fluoride
- KCl – potassium chloride
- Al_2O_3 – aluminum oxide

B. When multiple valence numbers or electrons are involved, we do the following for metals.

- Sn^{+4} = stannic
- Sn^{+2} = stannous

The larger number is -ic, the smaller is -ous.

Metals and oxygen: $Cu^{+2} + S^{-2} = Cu_2S_2$ or CuS (cupric sulfide)

$Cu^{+1} + O^{-2} = Cu_2O$ or cuprous oxide

BINARY COMPOUNDS

Chemists will often use prefixes when naming binary compounds (only two elements present)

- SO_3 (sulfur trioxide)
- CCl_4 (carbon tetrachloride)

Note: 3 is tri, 4 is tetra-, 5 is penta-, 6 is hexa-

- $AsCl_3$ (arsenic pentachloride)

RADICALS OR POLYATOMIC IONS

When radical decrease, we go from -ate to -ite.

- $Ca(NO_3)_2$ (calcium nitrate)
- $Ca(NO_2)_2$ (calcium nitrite)
- Hydrogen + Oxygen

As oxygen decreases, we go from -ic to -ous.

- H_2SO_4 (sulfuric acid)
- H_2SO_3 (sulfurous acid)

Note if oxygen increases, we use hyper.

- $HClO_4$ (hyperchloric acid)
- $HClO_3$ (chloric acid)

EXERCISE 21-1

Name the following compounds:

H_2O	Cl_2O_7	P_2O_3	NaOH
N_2O	H_2CO_3	CO_2	$Al_2\ (SO_3)^3$
HCl	Co	NaCl	H2SO4
$KClO_3$	H_2OSO_3	$Al_2\ (SO_4)_3$	PCl_5
KF	$AsCl_3$	HF	KOH
$HClO_4$	MgO	NaBr	

SPECIAL TOPICS 2

OXIDATION NUMBERS

The sum of oxidation numbers must equal zero. Examples:

+2 – 2 = 0 (+1) (-2)

H_2O +2 +6 -8 -6

H_2SO_4

Oxidation Numbers – a number assigned to an atom in a polyatomic ion or molecular compound. Hydrogen is +1 and -1 in a hydride. Group I is also +1. Oxygen is -2 and -1 in a peroxide. The sum of oxidation numbers must equal zero. We use algebra to calculate oxidation numbers.

Example: (+1)

+2 -2 -0

Note +2 and -2, -0

H_2 is +1 and oxygen is -2

Example: What is the oxidation number of S in H_2SO_4?

+2 +6 -8 = 0

H_2 S O_4 So sulfur = +6

+2 +12 -14

What is the oxidation number of chromium $K_2Cr_2O_2$?

Cr is +6 *Note we have 2Cr but one is +6*

What is the oxidation number of the Mn in $MnO\text{-}_4$?

The permanganate ion has a -1 charge, so we have

$\frac{+7\ -8}{MnO_4}$ because oxygen is -2 x +4 = -8 and +7 and -8 = -1

EXERCISE 21-2

Solve the oxidation numbers of the underlines

KCl	H_2SO_4	NaCl	K_2CrO_4	K_2CrO_7

Solve oxidation numbers of each element in the following.

SO_2	K_2SO_4	HNO_3	PO_4^{3-}
S_2O_3	N_2O_3	Nh_3	Na_2O_2

SPECIAL TOPICS 3

VOLUME AND MASS EQUATIONS

$2\ Mg + O_2 - 2\ Mg\ O$
8.4 liters of O_2 will yield ________ grams of MgO.
$2KClO_2 - 2KCl + 3\ O_2$
38 grams of $KClO_3$ ________ grams of KCl
108 grams of $KClO_3$ ________ liters of O_2
$N_{(g)} + 3H_{(g)} \rightarrow 2\ NH_{3(g)}$
12 liters of N_2 will yield ________ liters of NH_3
18 moles of N_2 will yield ________ moles of NH_3

SPECIAL TOPICS 4

EXERCISE 21-3

(Ksp Problems)

A. (1.) Convert to moles per liter and grams per liter. Ksp = 2.5×10^{-3}

$Ag + C_2\ H_3\ O_2\ [Ag^+]\ [C_2H_3O_2]$

(2.) Convert to grams per liter. (Refer to problem 1)

B. Convert to moles per liter. Ksp = 8.1×10^{-12}

$Ag_2CO_3\quad [Ag^+]^2\ [CO3^-]$

note: This problem = $(X)\ (2X)^2 = 4X^3$

C. A chemist dissolves 1.3×10^{-4} grams of AgBr in 1.0 liter. What is the Ksp of this solution?
D. Solve for Ksp of MgF2 with 7.4×10^{-2} grams per liter.
E. Solve for Ksp of $CaCO_3$ with 7.0×10^{-3} gram per liter.

SPECIAL TOPICS 5

RULES OF DULONG & PETIT

The Rule of Dulong and Petit refers to the study of specific heat and the atomic weights of the elements in their solid state. The formula is:

Atomic weight x Specific heat = 6.4

The figure 6.4 is an average of several elements and atomic heats.

We will use the Rule of Dulong and Petit to find the exact atomic weight of an element.

Procedure

1. Determine the equivalent weight of an element.
2. Calculate approximate atomic weight 6.4/specific heat.
3. Determine valence by dividing approximate weight by the equivalent weight.
4. Multiply the equivalent weight by the valence number = exact atomic weight.
5. Oxygen and hydrogen are often used. Equivalent atomic weight of oxygen is 8. Equivalent atomic weight of hydrogen is 1.

Example

A compound contains 52.9% Al and 47.1% O_2. The specific heat of Al is 0.215 cal/g°C. find the exact weight of hydrogen is 1.

1. Solve for equivalent wt of Al = 0.529/0.471 = X/8 = 8.98 grams of Al. Note 8 is atomic equivalent weight of oxygen.
2. Determine approximate weight:
   6.4/sp. ht = 6.4/0.215 = 29.7 grams for approximate weight.
3. Divide approximate weight by equivalent weight:
   29.7/8.98 = 3.3 round off answer to nearest whole number = 3 valence
4. Multiply the equivalent weight by valence = 3 x 8.98 = 26.94, therefore, 26.94 is the exact weight of Al.

Problem

A compound is 60% Mg and 40% O_2. The specific heat of Mg is 0.245cal/g°C.

SPECIAL TOPICS 6

MOLALITY AND COLLIGATIVE PROPERTIES

1. What is the weight of compound X if a technician dissolves 30 grams in 100 grams of water with a boiling point of 100.26°C.

   Solutions Note: 0.52°C + 100°C = 100.52°C = 1 molal based on H2O.

$$= \frac{M \ x\ 1\ Kg\ x\ b.p.constant}{1\ x\ ?Kg\ x\ ?b.p.} = \frac{30\ grams\ x\ 100\ grams\ x\ 0.52°C}{1\ x\ 100\ grams\ x\ 0.26°C} = 600\ grams$$

2. What is the weight of compound *X* is the solution freezes at -3.72°C after the chemist dissolved 50 grams in 250 grams of H_2O.

Solution: = mass x 1 kg x F.p. constant

Complete this chart: (page 180)

$C_6H_{12}O_6$ in 1 liter H_2O (A-C)

A. 0.6m	______ b.p	______ f.p.	______ grams in 1L
B. ______ m	100.30°C	______ f.p.	______ grams in 1Kg
C. ______ m	______ b.p.	-0.372°C f.p.	______ grams in 1L

A chemist dissolves 30 grams of compound X in 250 rams in H_2O. she records a b.p. of 100.26°C. What is the weight of this compound? What is the freezing point of this compound? Grams in1 liter or 1 Kg?

SPECIAL TOPICS 7

ISOTOPES

Isotopes are defined as elements with the same atomic number but a different atomic mass or weight.

Examples:

Oxygen or O has an atomic number of 16 and a mass of 32 (16 protons and 16 neutrons).

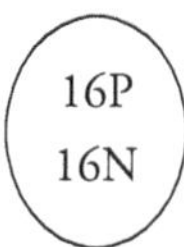

Isotopic oxygen would have an atomic number of 16 and a mass of 34.

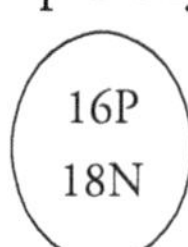

Note: We only change the number of neutrons and the atomic number or protons remains the same.

Further Study of Isotopes

Isotope Problem

The following information is hypothetical but the solution or concept is factual.

An element has two isotopes. One isotope has a relative abundance of 60% and a mass of 25 grams. The second isotope has a relative abundance of 40% and a mass of 22.5 grams. Find the weight of this element.

Solution

60% = 0.60 and 40% = 0.40 so we multiply the decimals by their mass and add the total sum for the weight of the element. Therefore:

0.60 x 25 grams = 15 grams
0.40 x 22.5 grams = 9 grams
15 grams + 9 grams = 24 grams

The mass of this element based on its isotopes is 24 grams or 24a.m.u. This element could be Mg, however, remember our data is hypothetical.

Isotopes: Example 2

Isotopes of Nitrogen

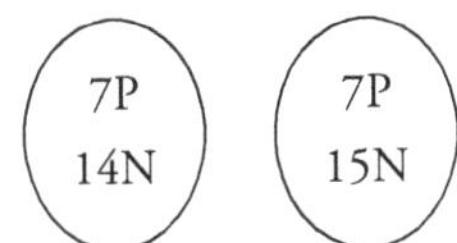

14.7 N 99.63% abundance
15.7 N 0.37% abundance

Note: 14 and 15 are different weights and 7 is the atomic number so it *does not change*

Find the atomic weight of Nitrogen

Solution 14 x 0.9963 = 13.9482
15 x 0.0037 = 0.055
13.9482 + 0.055 = 14 grams

The average atomic mass of nitrogen is 14 a.m.u

SPECIAL TOPICS 8

(PERCENT YIELD)

The percent yield is equal to actual yield divided by theoretical yield

The actual yield is what you get from the actual lab reaction. On the other hand. The theoritecal yield is based on desk or lecture data. Chemists calculate percent yields because they compare the actual lab results to lecture or non-lab data.

Same Problem:

The explosive TNT is made by reacting toluene C_7H_8 with nitric acid HNO_{3}.

$$C_7H_8\ (L) + 3\ HNO(aq) \rightarrow C_7H_5(NO_2)_3 + 3\ H_2O$$

A chemist's has the following lab data.
38 grams of C_7H_8 (L) yielded 90 grams of TNT $C_7H_5(NO_2)_3$. Find the % yield.

Solution:
The key to this problem is to divide actual empirical results to theoretical yield.

<u>**Problem 2**</u>
$Mg + 2\ HCl \rightarrow MgCl_2 + H_2$
A lab technician used 21 grams of Mg and records 18.5 of H gas. Find the % purity.
What is the percent purity if the following equation:
$CaCO_2 \rightarrow CaO + CO_2$
If a chemist's use 30 grams of $CaCO_3$ and his lab results yield 16 grams of CaO.

Solution:
Solve for theoretical data (yield) then compare theoretical to lab results (actual yield).

$$percent\ purity = \frac{lab\ results}{lecture\ results}$$

$$30\ g\ CaCO(s)\ x\ \frac{1\ \cancel{mole}}{100\ g\ CaCO_{3(g)}}\ x\ \frac{56\ f\ of\ CaO(s)}{1\ \cancel{mole}} = 16.8\ grams\ of\ CaO(s)$$

$$percent\ purity = \frac{lab\ results}{lecture\ data} = \frac{16\ g\ CaO(s)}{16.8\ g\ CaO(s)} = 95.2\%$$

SPECIAL TOPICS 8

(PERCENT YIELD)

<u>Actual</u>

The lab results yielded 90 grams of TNT. Since we have actual we must compare to the theoretical.

Theoretical data or lecture data is: $38\ g\ C_7C_8\ x\ \frac{1\ mole\ C_7C_8}{92\ grams}\ x\ \frac{227\ g}{1\ mole\ TNT} = 93.76\ grmas\ of\ TNT$

The percent yield is $\frac{lab\ result}{lecture} = \frac{90\ g\ TNT}{93.76\ TNT}\ x\ 100 = 96\%$

Percent yield is 96%

The chemist's lab work yielded 96% of the predicted or lecture data and calculations.

Percent Yield Problems

Solve for % yield based on the following equation.

$2\ KClO_3 \rightarrow 2\ KCl + 3\ O_2$

A technician burns 40 grams of $KClO_3$ and records 19.8 grams of KCl. What is the percent yield?

SPECIAL TOPICS 9

(ENERGY)

Calculate H or change in enthalpy.

$Cl_{2(g)} + 2HBr_{(g)} \rightarrow 2HCl + Br_{2(g)}$

First we refer to our chart thermodynamic value and find ΔH HBr = -36.4 kJ/mol, ΔH for HCl = -92.3 kJ/mol

Note: Br_2 and $Cl_2 = 0$

ΔH° = E ΔHf (reactants)

ΔH° = products (2HCl = -92.3 x 2 = -184.6 kJ/mole)

ΔH reactants (2HBr = -36.4 x 2 = -72.8 kJ/mol)

-184.6 kJ/mol + 72.8 = -111.8 kJ

ΔS° = change in entropy

Find ΔS based on the following equation

$CaCO_{3(g)} \rightarrow CaO_{(g)} + CO_{2(g)}$

E S° products – E S° reactants

$CaCO_3$ = -92.9 J/k

CaO = 38.2 J/k

$CO_{2(g)}$ = 213.74 J/k

Products – Reactants

$CaO + CO_{2(g)} \rightarrow CaO_{3(g)}$

38 .2 J/k + 213.74 J/k – (-929.9 J/k)

= 251. 94 J/k + 92.9 J/k

= 344.84 J/k

ENERGY, ENTHALPY AND ENTROPY

TERMS

- Energy – the ability to do work
- Thermodynamics – the study of matter and energy interactions
- G or Gibbs – free energy
- H or enthalpy or heat content
- ΔS = entropy or disorder
- exothermic – reaction = loss of heat or a negative delta G
- endothermic reaction – heat is absorbed or + G

THERMODYNAMIC CHART

Enthalpies o Formation (ΔH_f) in kJ $mole^{-1}$ at 25°C and 1 atm.

Al_2O_3 (s)	-1670	HF (g)	-286.6
$BaCO_3$ (s)	-1217	HI (g)	25.9
BaO (s)	-558.6	HCOOH (l)	-409.2
$CaCO_3$ (s)	-1207	HgO (s)	90.71
CaO (s)	-635.1	HNO_3 (aq)	-207
CH_4 (g) (methane)	-74.85	H_2O (g)	-241.8
CH_3OH (l) (methanol)	-238.6	H_2O (l)	-285.8
C_2H_2 (g) (acetylene)	226.7	H_2O_2 (l)	-187
C_2H_4 (g) (ethylene)	52.30	H_2S (g)	-20.12
C_2H_5OH (l) (ethanol)	-277.7	H_2SO_4 (aq)	-907.5
C_2H_6 (g) (ethane)	-87.68	NaCl (s)	-410.9
C^6H^6 (l) (benzene)	49.04	NH_3 (g)	-46.19
$C^6H_{12}O^6$ (s) (glucose)	-1260	NO (g)	90.37
$C_{12}H_{22}O_{11}$ (s) (sucrose)	-2221	NO_2 (g)	33.85
CO (g)	-110.5	N_2O (g)	81.55
CO_2 (g)	-393.5	SO_2	-296.9
Fe_2O_3 (s)	-822.2	SO_3	-395.2
HBr (g)	-36.23	ZnO (s)	-83.2
HCl (g)	-92.30	ZnS	-48.5

ΔG = Gibbs free energy
$\Delta G = E\ \Delta G°f$ (products) – $E\ \Delta G°f$ (reactants)
$2H_2O_2 \rightarrow 2H_2O\ L + O_{2(g)}$

From table

H_2O_2 = -120 kJ/mol
H_2O = -237 kJ/mol

Note: Oxygen = zero J/k

$\Delta 6 = E\ \Delta G = f$ (products) – $E\ \Delta G$ (reactants)

$2H_2O\ L + O_2 \rightarrow 2H_2O_2\ L$
$\Delta G°$ = products
2 (-237 KJ) = -474

$\Delta G°$ = reactants
2 (-120 KJ) = -240 KJ

Since we subtract -476 + 240 = -234.7 KJ

ΔG = -234.7 KJ

Williard Gibbs is an American chemist developed an equation for entropy, enthalpy and free energy to be.

$\Delta G° = \Delta H° - T\ \Delta S$

Or free energy = a chane in enthalpy minus temperature multiplied by entropy.

Since $\Delta G° = \Delta H° - T\ \Delta S$

Note: T = Kelvin temperature or 273°K
ΔS = change in entropy
ΔH = change in enthalpy

<u>**EXAMPLE**</u> Find ΔG
A technician recorded a temperature of 60°C. A **ΔS** of -0.09 kcal/K and a **ΔH** of -87 kcal.

Solution: Since $\Delta G = \Delta H - T\ \Delta S$ = -87 kcal – (333°K x -0.09 kcal/K)
= -87 kcal - (-29.97)
= -57 kcal

Note: a negative ΔG is a spontaneous reaction.

SPECIAL TOPICS 10

(CALORIMETER PROBLEM)

A calorimeter is used to measure the amount of heat absorbed or released in a chemical reaction. The temperature change of the water is used to determine the amount of heat lost or gained.

heat lost = heat gained

CALORIEMETER PROBLEM

A chemist's drops a piece of Al with a mass of 18.5 grams at a temperature of 82.5°C in an insulated cup of water. The water has a mass of 135 grams and its temperature before adding the Al was 17°C. Find the final temperature of the system (the specific heat of Al is 0.2159 cal g/°C).

Solution: Mass x Specific heat x Change in temperature
Al = 18.5 grams x 0.2159 x 65.5°C = 261.6 cal g/°C
Water = 135 x 1 = 135
Al = 261.6 and H_2O = 135 = 261.6 / 135 = 1.93°C

Since the initial temperature was 17 we add the new temperature or 1.93°C + 17°C = 18.93°C or 19°C.

Our final temperature is 19°C.

SPECIAL TOPICS 11

(THE LAW OF HESS)

Some chemical reactions are difficult to measure. This is so because some are explosive, have side reactions and others are extremely slow. The Law of Hess enables us to measure these difficult reactions in the lab.

Law of Hess Problem

A. $CuO + H_{2(g)} \rightarrow Cu_{(s)} + H_2O_{(g)}$ Find ΔH
We note that equation A is the basis for solving this problem or the main equation.
Solution $\Delta H_1 + \Delta H_2$

The following is based on a table listing heats of formation.

B. $H_2 + \frac{1}{2} O_2 \rightarrow H_2O$ ΔH = -57.8 Kcal
C. $CuO_{(g)} \rightarrow Cu_{(g)} + \frac{1}{2} O_2$ ΔH = -37.1 Kcal

Note: Equation B is direct combination as is equation A. (Study H_2)

On the other hand equation C differs from equation A because A shows CuO as decomposition, but in C. Cu is direct combination.

Therefore, we must change the sign of Cu in equation C. From -37.1 Kcal to +37.1 Kcal.

Therefore: -57.8 Kcal
+37.1 Kcal
-27.7 Kcal

SPECIAL TOPICS 12

(THE NERNST EQUATION)

The Nernst Equation is often used to solve voltage when the concentrations are not standard or 1.0 M.

Summary of Nernst Equations

A. The greater the concentration of reactants the more spontaneous the reaction or a positive voltage.
B. On the other hand a negative voltage or less spontaneous reaction occurs if the product concentration increases or a recent concentration or decrease.
C. The Nernst equation is used to calculate the effect of concentration or molarity upon voltage.
D. The concentration should be less than one or more than one.

The Nernst Equation

(at 25°C and in terms of base ten logarithms) is $E° = \frac{E° - 0.0592 \log Q}{n}$

Q is the quotient
E id the voltage
N is the number of oxidation electrons

Calculate E cell at 25°C based on the following equation

$$2Al_{(g)} + 3Mn^{2+}(aq) = 2Al^{3+}(aq) + 3Mn_{(s)}$$

Where E cell = 0.75 volts
Concentrations are (Mn2+) = 0.70M
(Al3+) = 1.75M

Note: Al is oxidized form zero to 3^+ and Mn^{2+} is reduced from 2^+ to zero.

We use the oxidized elements Al, which is increased by 3

n = 6 ($2\,Al \rightarrow 2\,Al^{3+} + 6e^-$)

($3\,Mn^{2+} + 6e^- \rightarrow 3Mn$)

or multiply oxidation state by reduction

n = Al 3 x Mn 2 = 6

E° = E° - 0.0592/n log Q

E° = 0.75 volts

$$Q = \frac{(Al^{3+})^2}{(Mn^{2+})^3} = \frac{1.75^2}{0.70^3} = \frac{3.0625}{0.343} = 8.928$$

or

Log of 8.928 = 0.95 final solution

$$0.75 - \frac{0.0592}{6}\ x\ 0.95 = 0.75 - 0.0094 = 0.74\ volts$$

E° = 0.74 volts

CHAPTER 22

A FEW NEW ACHIEVEMENTS IN CHEMISTRY

ENERGY

- Oil – basis of petroleum, gasoline, energy
- Lights – increase visibility and energy
- Coal – energy
- Lasers – can be used for surgery, communications
- Electric – redox reaction
- Nuclear – study of the nucleus, energy, weapons, radiation
- Solar – the rays of the sun or light, windmills, energy
- Gasohol – alcohol and gasoline
- Water – amazing compound with H bonding, universal solvent
- Sunflowers – can be used for energy
- Water – can be recycled for industry and energy

COMMUNICATINS AND ELECTRONICS

- Crystals – used in calculators, televisions, solid state physics
- Computers – countless uses
- Radios, televisions, telephone – involves most fields of science

HELPFUL CHEMICAL PRODUCTS

- Plastics – explosives, containers, bottles, gloves, plates, bags utensils
- Synthetic clothing – polyester
- Synthetic diamonds – cubic zirconium
- Bullet-proof glass, fireproof clothes (firemen, baby blankets) – all uses of silicon compounds

CHAPTER 23

ORGANIC CHEMISTRY

CHARACTERISTICS OF ORGANIC COMPOUNDS

1. Generally organic compounds are non-polar and insoluble in water; water is polar.
2. Some organic compounds will dissolve in non-polar solvents such as alcohols, benzene and gasoline.
3. Most organic compounds are non-electrolytes.
4. Organic acids are exceptions because they are weak electrolytes.
5. Low melting points.
6. Covalent bonding.
7. Slow reactions.

SOURCES

1. Plants
2. Animals
3. Natural gas
4. Coal
5. Petroleum

CATENATION

Carbon can form countless compounds because of its ability to join or combine with other carbon atoms.

ORGANIC COMPOUNDS

1. Alkanes – all single bonds
2. Alkenes – at least one double bond
3. Alkynes – at least on triple bond

GENERAL SURVEY OF ORGANIC COMPOUNDS

Figure 23-1

Alkane — single bonds
Alkene — double bonds
Alkyne — triple bonds
Aromatic — rings such as benzene

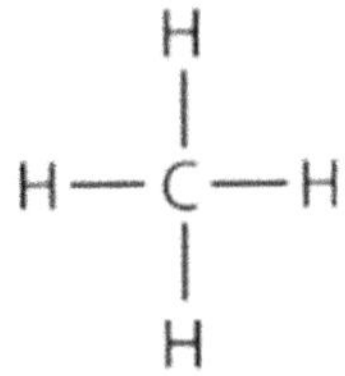

```
    H
    |
H — C — H
    |
    H
```

GENERAL SURVEY OF ORGANIC COMPOUNDS
Figure 23-2

HYDROCARBON	STRUCTUAL FORMULA	MODEL
Methane Ch_4	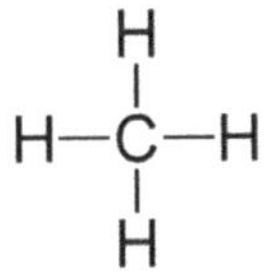 H H–C–H H	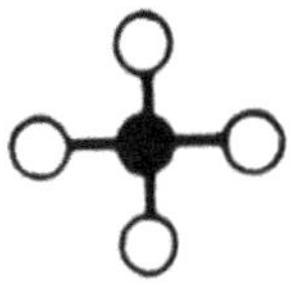
Ethane C_2H_6	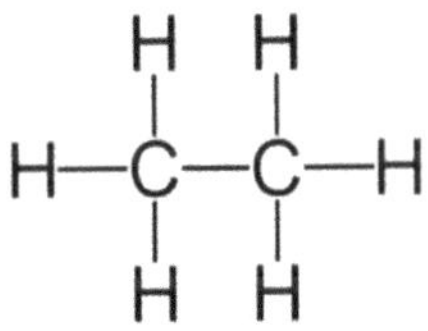 H H H–C–C–H H H	
Propane C_3H_8	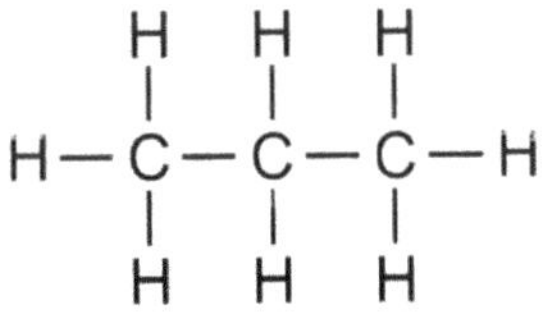 H H H H–C–C–C–H H H H	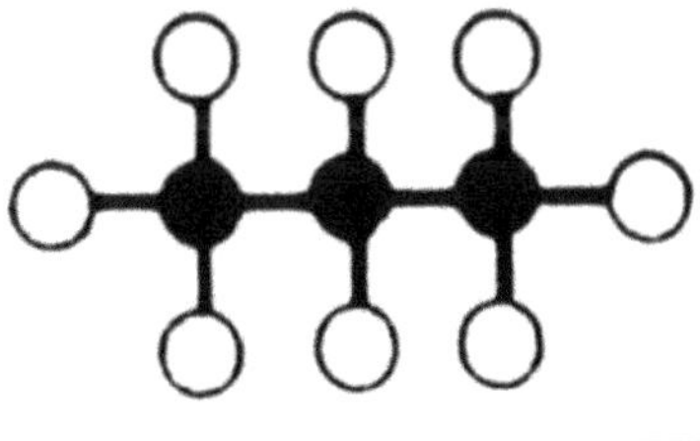
Butane C_4H_{10}	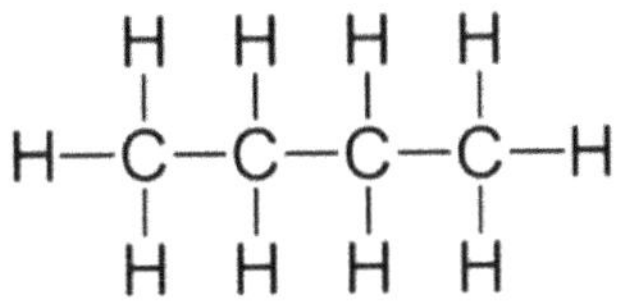 H H H H H–C–C–C–C–H H H H H	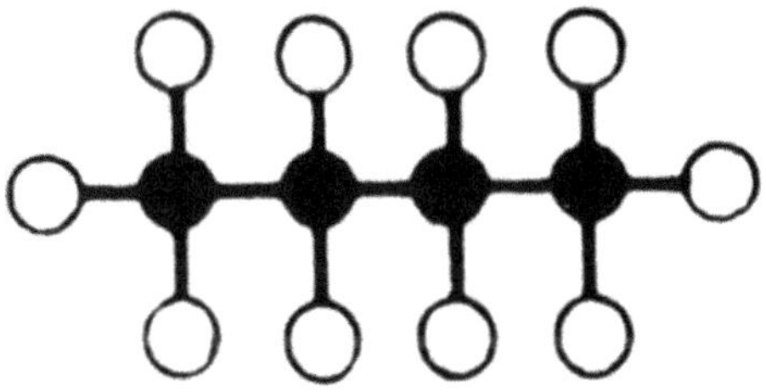
Pentane C_5H_{12}	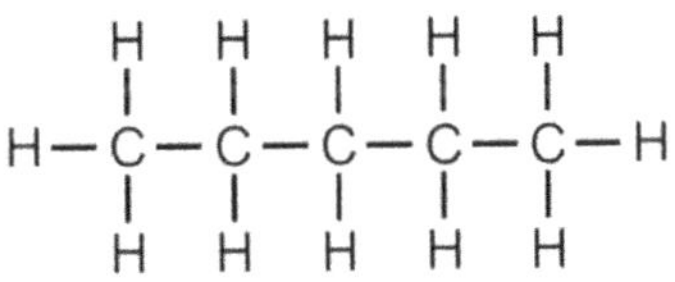 H H H H H H–C–C–C–C–C–H H H H H H	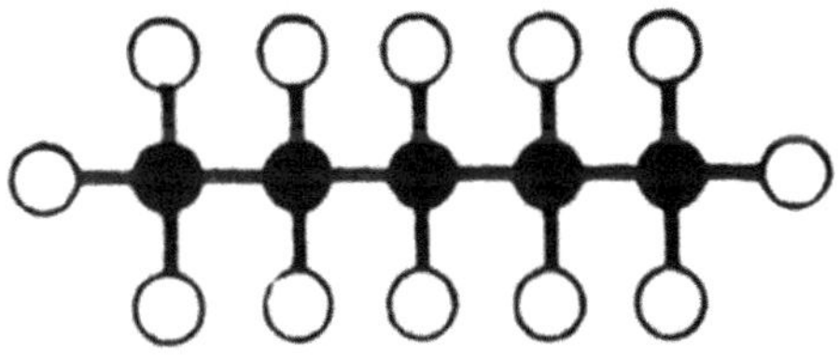

Figure 23-3

Properties of the alkanes and alkyl radicals.

ALKANE	MOLCULAR FORMULA	MELTING POINT °C	BOILING POINT °C	STATE (20°C)	NAME OF ALKYL RADICAL	FORMULA OF RADICAL
Methane	CH_4	-183	-161	Gas	Methyl	CH_{3-}
Ethane	C_2H_6	-172	-89	Gas	Ethyl	C_2H_{5-}
Propane	C_3H_8	-187	-42	Gas	Propyl	C_3H_{7-}
Butane	C_4H_{10}	-135	-0.5	Gas	Butyl	C_4H_{9-}
Pentane	C_5H_{12}	-130	36	Liquid	Pentyl	C_5H_{11-}
Hexane	C_6H_{14}	-95	69	Liquid	Hexyl	C_6H_{13-}
Heptane	C_7H_{16}	-91	98	Liquid	Heptyl	C_7H_{15-}
Octane	C_8H_{18}	-57	126	Liquid	Octyl	C_8H_{17-}
Nonane	C_9H_{20}	-54	151	Liquid	Nonyl	C_9H_{19}
Decane	$C_{10}H_{22}$	-30	174	Liquid	Decyl	$C_{10}H_{21-}$

Figure 23-4

Some members of the alkene series.

NAME	STRUCTURAL FORMULA	MODEL
Ethene (Ethylene) C_2H_4	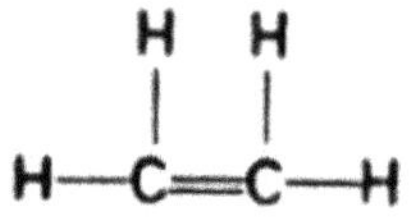	
Propane (Propylene) C_3H_6	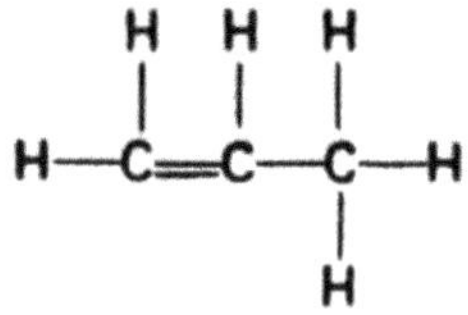	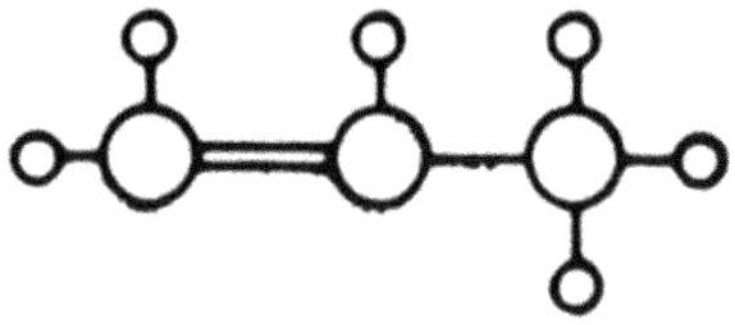
Butene (Butylene) C_4H_8	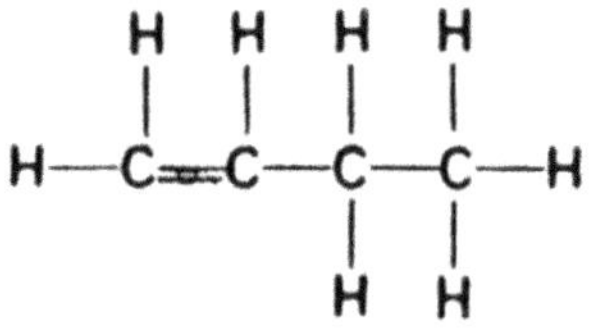	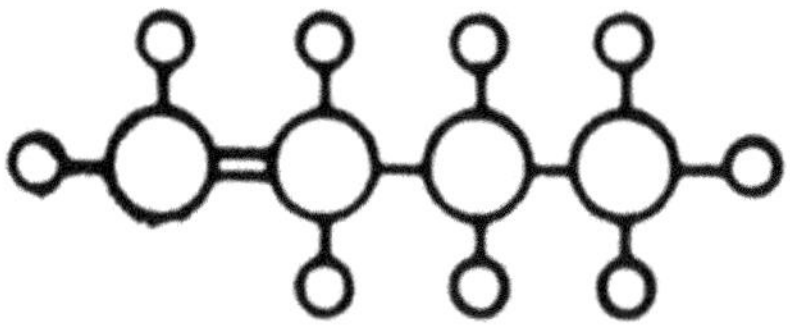
Pentene C_5H_{10}	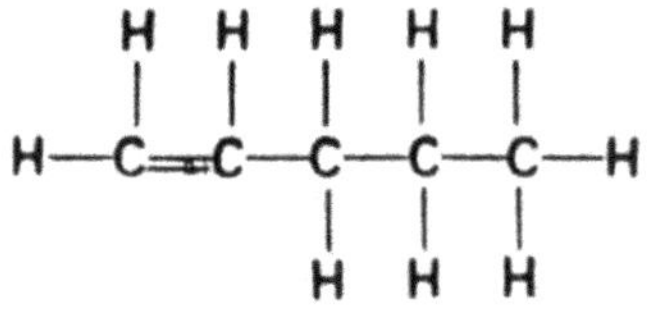	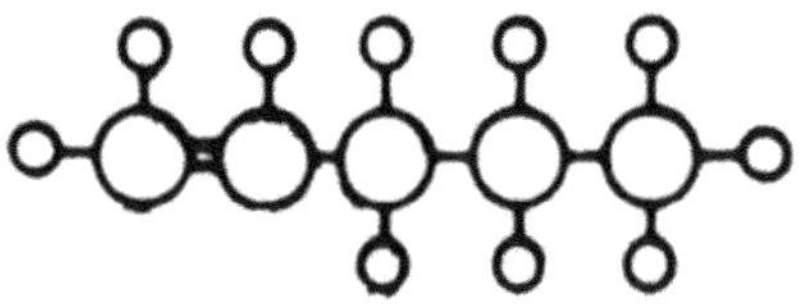

Alkene series. One common alkadiene used in the production of synthetic rubber is 1, 3-butadiene.

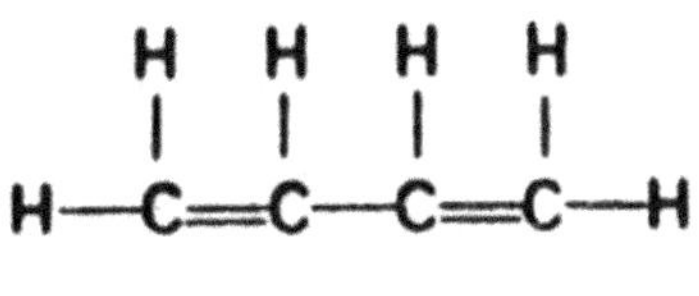

butadiene

Figure 23-5
Classification of hydrocarbons.

	TYPE	NAME	EXAMPLE	STRUCTURAL FORMULA	GENERAL FORMULA OF SERIES
Alphatic	a. Saturated	Alk*anes* Have only single bonds	Eth*ane* C2H6	H H H—C—C—H H H	C_nH_{2n+2}
	b. Unsaturated	Alk*enes* Contain one double bond	Eth*ene* C2H4	H H H—C═C—H	C_nH_{2n}
		Alk*ynes* Contain one triple bond	Eth*yne* C2H2	H—C≡C—H	C_nH_{2n-2}
Aromatic		Ar*enes*	Benzene C6H6	H C H—C C—H H—C C—H C H	C_nH_{2n-6}

Figure 23-6
Organic functional groups.

NAME	GROUP	GENERAL FORMULA	EXAMPLE	NAME
Alcohol	$—OH$	$R—OH$	CH_3CH_2OH	Ethanol (ethyl alcohol)
Aldehyde	$—C(=O)—H$	$R—C(=O)—H$	CH_3CHO	Ethanal (acetaldehyde)
Ketone	$—C(=O)—$	$R—C(=O)—R^1$	CH_3COCH_3	Propanone (acetone)
Ether	$—O—$	$R—O—R^1$	$CH_3CH_2OCH_2CH_3$	Ethoxy ethane (diethyl ether)
Acid	$—C(=O)—OH$	$R—C(=O)—OH$	CH_3COOH	Ethanoic acid (acetic acid)
Halide	$—X$ (Br, Cl, I)	$R—X$	C_2H_5I	Iodoethane (ethyl iodide)
Ester	$—C(=O)—O$	$R—C(=O)—O—R^1$	CH_3COOCH_3	Methyl methanoate (methyl acetate)
Amine	$—NH_2$	$R—NH_2$	CH_3NH_2	Aminomethane (methylamine)
Amide	$—C(=O)—NH_2$	$R—C(=O)—NH_2$	CH_3CONH_2	Ethanamide (acetamide)
Amino Acid	$—C(=H)(NH_2)—C(=O)OH$	$R—C(=R^1)(NH_2)—C(=O)OH$	NH_2CH_2COOH	Glycine

Figure 23-7

	Structure	Monomer	Polymer	Uses
These polymers were essentially unknown before WW II	$H_2C{=}CH_2$	Ethylene	Polyethylene	Bags, coating, toys
	$H_2C{=}CH(CH_3)$	Propylene	Polypropylene	Beakers, milk cartons
	$H_2C{=}CHCl$	Vinyl chloride	Polyvinyl chloride, PVC	Floor tile, raincoats, pipe. Phonograph records
	$H_2C{=}CH(Cn)$	Acrylonitrile	Polyacrylonitrile, PAN	Rugs; Orlon and Acrilan are copolymers with other monomers
	$H_2C{=}CH(C_6H_5)$	Styrene	Polystyrene	Cast articles using a transparent plastic
	$H_2C{=}C(CH_3)C({=}O){-}OCH_3$	Methyl methacrylate	Plexiglas, Lucite, acrylic resins	High quality transparent objects, latex paints
	$F_2C{=}CF_2$	Tetrafluoroethylene	Teflon	Gaskets, insulation, bearings, pan coatings

QUESTIONS

1. Describe some characteristics of organic compounds.
2. How do alkanes, alkenes and alkynes differ?
3. List some modern uses of organic compounds.

CHAPTER 24

BIOCHEMISTRY

Biochemistry – is chemistry in living things.

Selected Example: Carbohydrates C, H, O in a 1:2:1 ratio. Starches and sugars are carbohydrates. Most carbohydrates are rich in energy.

Examples: table sugar or sucrose $C_{12}H_{22}O_{11}$
grape sugar or glucose $C_6H_{12}O_6$

Proteins – body builders, found in body tissues (blood, skin, eyes, hair, etc)
– are giant molecules based on amino acids, several amino acids equal a protein

Basic Amino Acids Structure

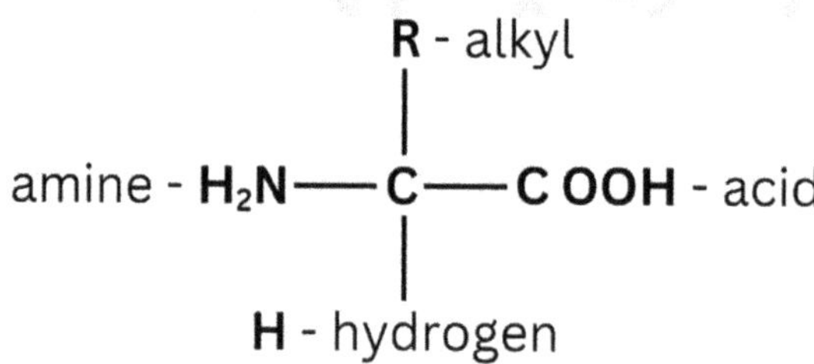

FERMENTATION is a chemical process whereby molecule are broken down. Enzymes, secreted by living organisms, act as catalysts in fermentation. Yeasts, for example, secrete enzymes that convert glucose into ethanol and carbon dioxide. The enzymes secreted by yeasts in the fermentation of sugar are collectively known as zymase.

$$C_6H_{12}O_6 \xrightarrow{\text{zymase}} 2C_2H_5OH + 2CO_2$$

Selected Compounds

ATP = energy
DNA = heredity
C = presentation in all living things

DNA Structure

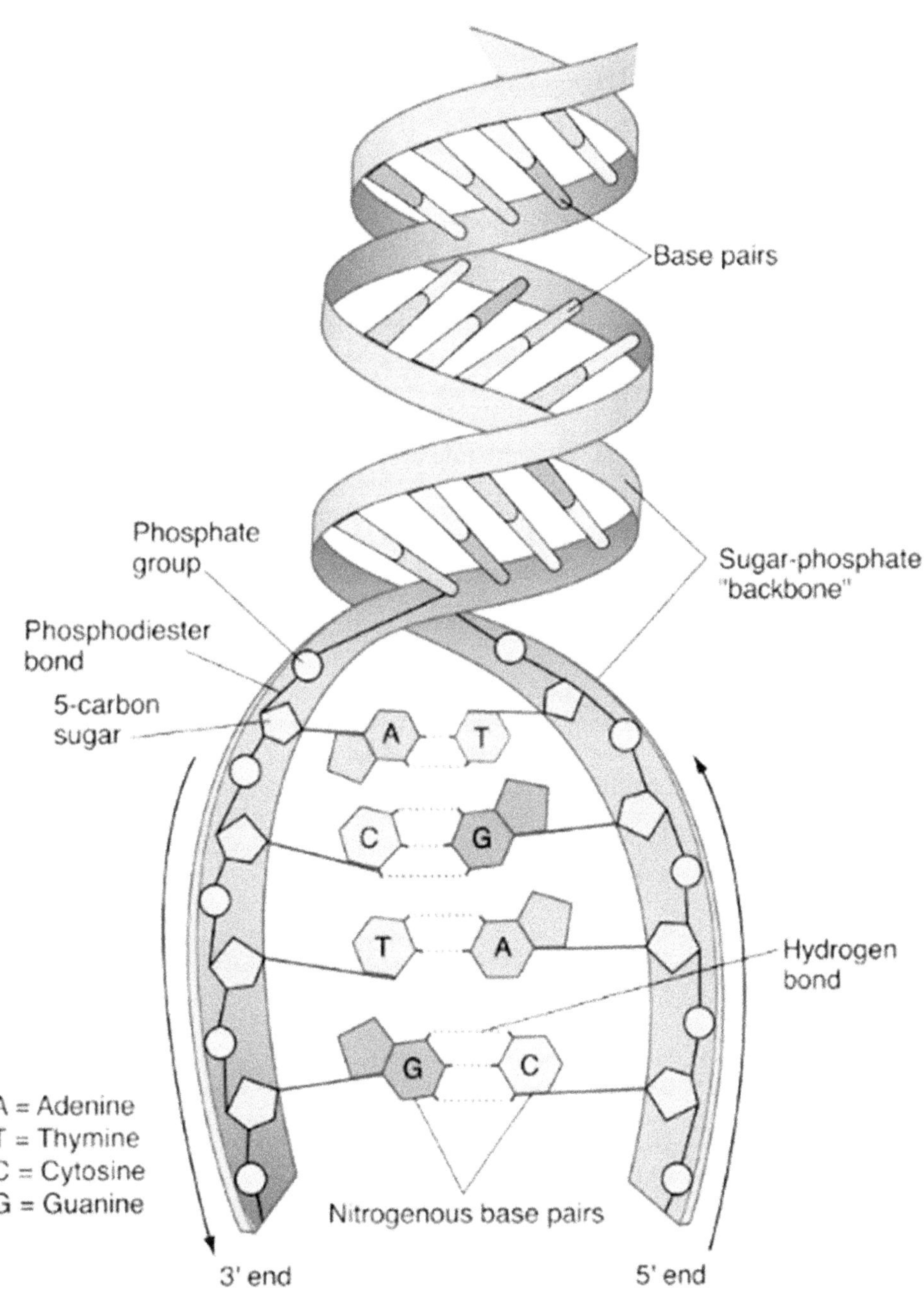

NATURALLY OCCURRING CARBOXYLIC ACIDS

NAME	STRUCTURE	NATURAL SOURCE
Acetic acid	$CH_3 — COOH$	Vinegar
Citric acid	OH | $HOOC — CH_2 — C — COOH — CH_2$ | COOH	Citrus fruits
Lactic acid	$CH_3 — CH — COOH$ | OH	Sour milk
Malic acid	$HOOC — CH_2 — CH — COOH$ | OH	Apples
Oleic acid	$CH_3(CH_2)_7 — CH = CH — (CH_2)_7 — COOH$	Vegetable oils
Oxalic acid	$HOOC — COOH$	Rhubarb, spinach, cabbage, tomatoes
Stearic acid	$CH_3(CH_2)_{16} — COOH$	Animal fats
Tartaric acid	$HOOC — CH — CH — COOH$ | | OH OH	Grape juice, wine

Organic acids are present in many foods.

ther, like aspirin and Vitamin C are more complex

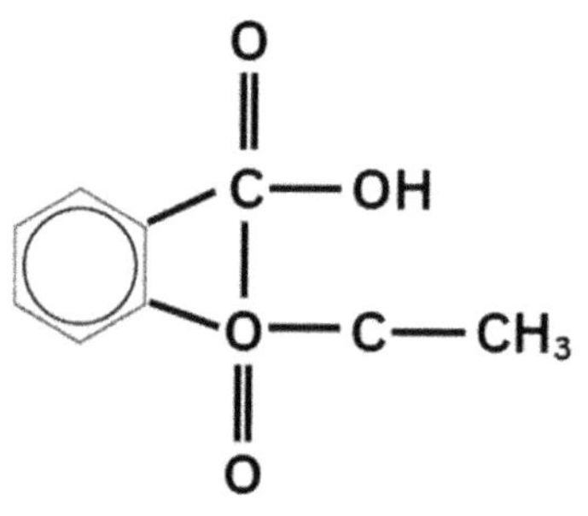

O
‖
C—OH
|
O—C—CH_3
‖
O

acetylsalicylic acid (aspirin)

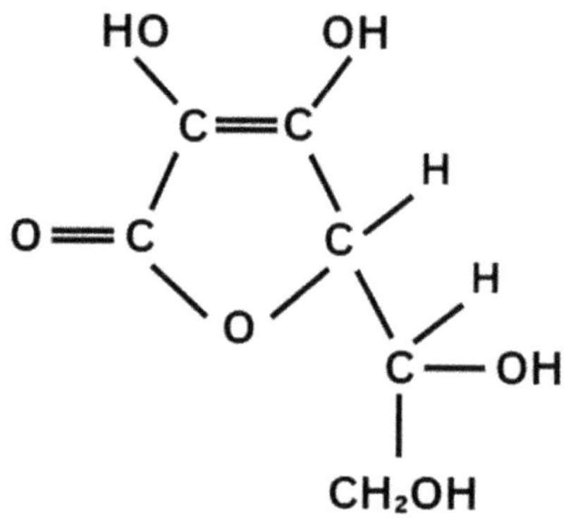

HO OH
C=C
H
O=C C
H
O
C—OH
|
CH_2OH

ascorbic acid (Vitaminc C)

Questions

1. Define carbohydrate, proteins and enzymes.

CHAPTER 25

Library Notes

Goals and Objectives

To be able to:

- Encourage the student to study various chemistry topics
- give the student an idea of some branches of chemistry

Key words

- biochemistry
- organic chemistry
- atomic research
- bonding
- formulas
- ideal gas laws

CHAPTER 25

LIBRARY TOPICS

KINETICS
EQUILIBRIUM
BONDING
ACIDS AND BASES

KINETICS AND EQUILIBRIUM ESSAYS

Activation Energy

Activation energy is the energy required to initiate a chemical reaction.

In order for a reaction occur colliding of molecules bring enough energy to the collision for the rearrangement of atoms to form new molecules. These collisions must be great enough to overcome an energy barrier or threshold. Several factors such as catalyst, temperature, concentration, nature of reactant and pressure can influence activation energy.

Moreover a pool player must hit the ball hard enough to cause collision or ball movement. One could compare pool ball movement to activation energy collisions. A minimum amount of energy is needed for the pool balls to collide and move.

In the top sequence, the bowler did not give the ball enough energy to get over the energy barrier. In the bottom sequence, the ball has enough energy to go over the energy barrier.

The Activated Complex

The activated complex is the arrangement of atoms at the peak of the activation energy barrier. This complex is unstable because atoms could become an ionic or molecular product. Furthermore this complex is short lived about 10-13 seconds and could form reactants or products. Since the *activated complex* could go either way it is also called the *transition state.*

(A) Two reacting particles approach each other. (B) They form an activated complex. (C) After forming the activated complex, the complex may break apart to reform the original substances, or (D) it may break apart to form one or more new substances.

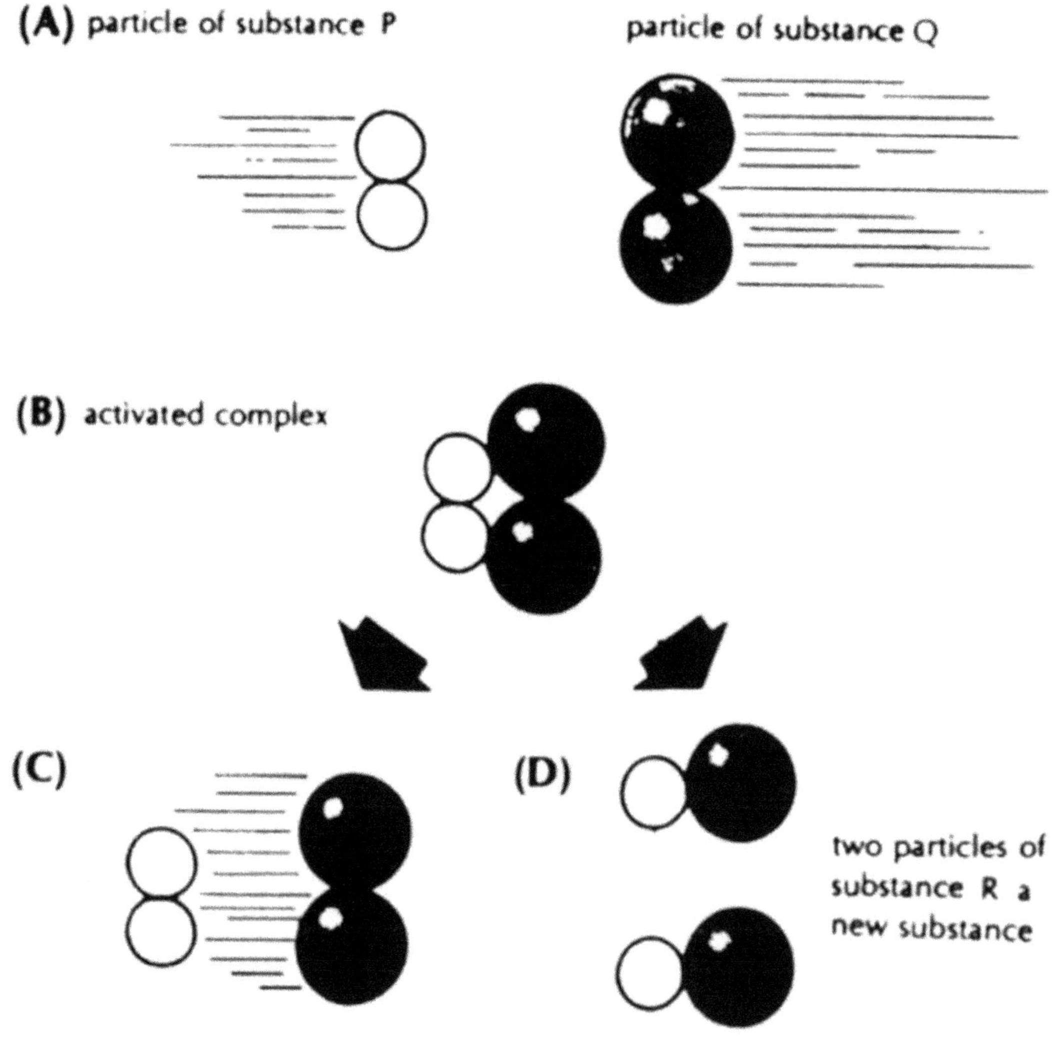

The Catalyst

A catalyst increases the rate of a chemical reaction without being used up itself in the reaction.

Simplified: A catalyst increases the rate of a chemical reaction by lowering the activation energy of the reaction.

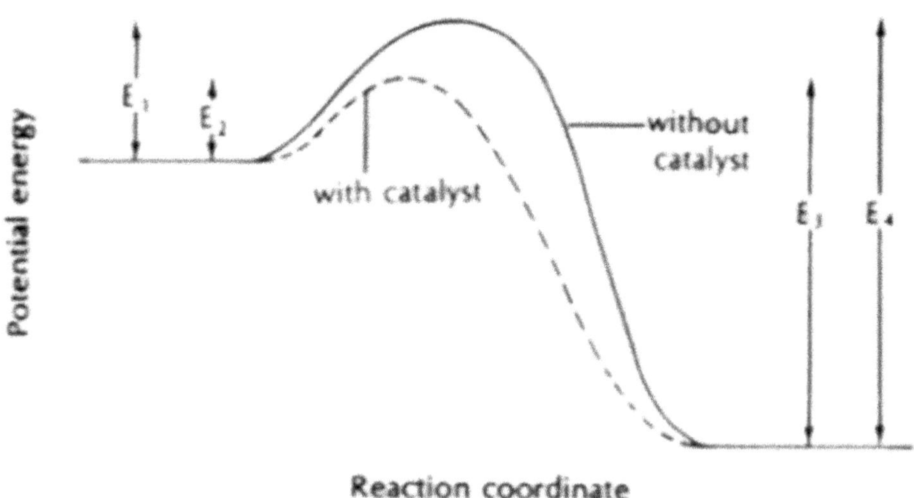

Effect of a catalyst. A catalyst reduces the activation energy required for a given reaction, thus making it possible for the reaction to proceed at a lower temperature or at a faster rate. The net energy change of the reaction is not affected. That is, the difference between the energy of the reactants and the energy of the products remains unchanged.

Exothermic Reactions

In exothermic reactions the products have less energy than the reactants. Energy is given off. The enthalpy or heat gained is negative. This is written as ΔG- (negative dela G)

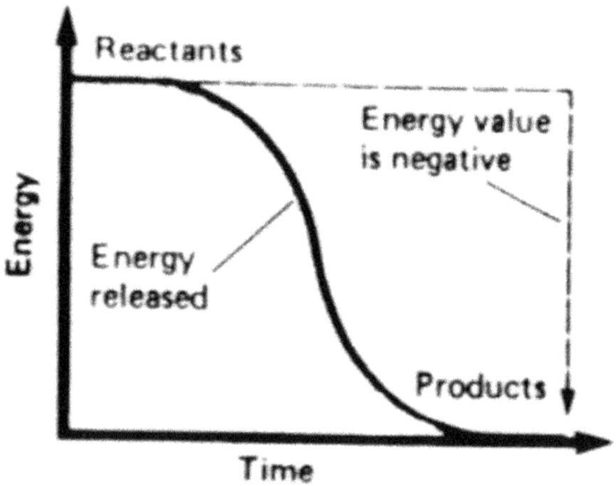

Products of another exothermic reaction are at a lower energy state than the reactants that lose some energy to the surroundings.

Endothermic Reactions

In the endothermic reactions the products have more energy than the reactants. The delta G or enthalpy is positive (ΔG+). Note: heat or energy is absorbed in endothermic reactions.

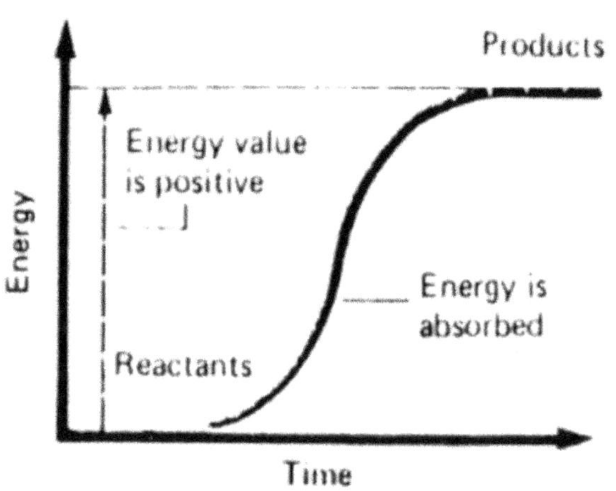

Energy is absorbed during an endothermic reaction so products are at a higher energy state.

Quantum Mechanics

Quantum mechanics incorporated both particle wave theory and probability into its description of the energy of an electron in the atom.

Key Terms

- Orbital – region around the nucleus where an electron is most likely to be found
- Ground state – when hydrogen atoms are in their lowest energy state
- Atomic number – the number of protons in an element, often written as Z
- Principal quantum number – describes the electron in an orbital
- Energy levels – hydrogen atoms can only exist in definite energy levels

Orbitals of the hydrogen atom

energy levels	number of orbitals
(n)	n^2
1	2
2	4
3	9
4	16

Simplified: The n^2 determines the number of orbitals. 1 x = 1 or 1^2, 2^2 = 4 orbitals.

***Note:* Pauli Exclusion principle states that electrons spin in opposite directions and 2 electrons equal one orbital.**

The first four principal energy levels and their sublevels. Note that the 3d sublevel has higher energy than the 4s, producing an overlapping of principal energy levels 3 and 4. Within a given principal energy level, *s* sublevels are lower in energy than *p* sublevels, *p* sublevels are lower than *d* sublevels, and *d* sublevels are lower than *f* sublevels. This diagram should be compared closely with the preceding figure.

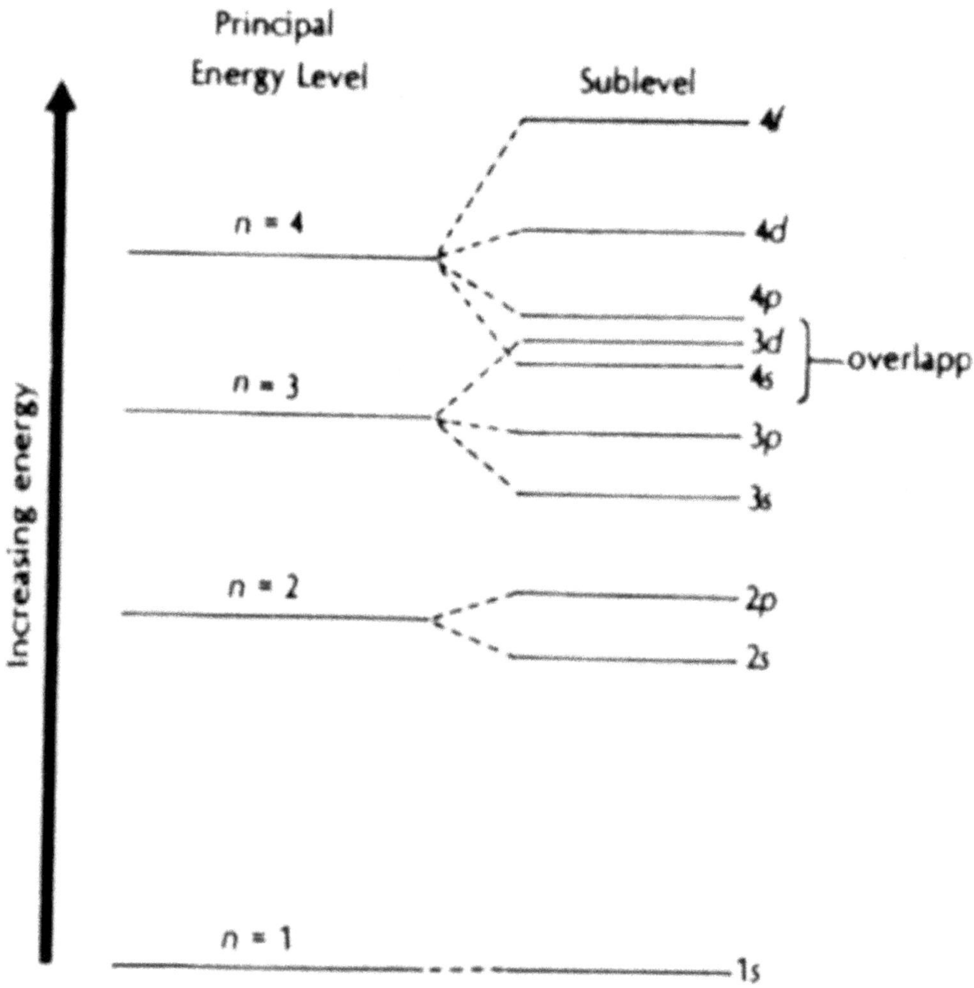

Therefore, 1 orbital = a maximum of two electrons

How many electrons in the 3rd quantum energy level?

Solution n^3 = 3 x 3 = 9 orbitals.
multiply 9 orbitals by 2 electrons (the maximum per orbital) = 18 electrons

Solve the following:
n^2 Find the number of orbitals
Solution 2^2 = 4 orbitals
electrons = orbitals x 2e- = 8 electrons

For further study:
Write a library report on quantum numbers

Summary of quantum numbers

- 1 = size of electrons
- 2 = shape of orbitals (s,p,d,f)
- 3 = axes (x, y and z axes)
- 4 = spin of electrons (Pauli exclusion principal)

Library Topic

Write a library report on the VSEPR theory (Valence Shell Electron Pair Repulsion)
This theory describes the equal distribution of electron pairs around a central atom in a molecule.

Bonding Review

Type of bond	Traits	Examples
Covalent bonding	Atoms share a pair of electrons	H_2, F_2, gases
Polar covalent	Unequal sharing of electrons	HF, HCl
Ionic bonds	Bonding between metallic and nonmetallic elements when + ion is attracted to the – ion	NaCl KCl
Hydrogen bonding	The weak attraction of a positive hydrogen atom in a molecule to an electronegative atom of another similar molecule	H_2O NH_3
Metallic bonding	Bonds with closely spaced energy levels that permit electrons to shift easily in metallic solids	Cu

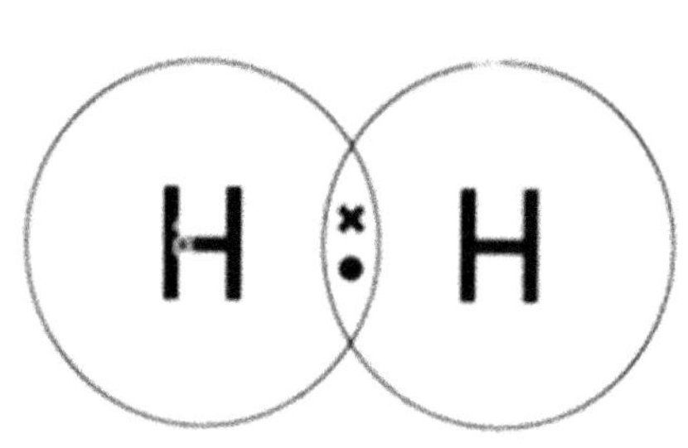

Hydrogen Molecule

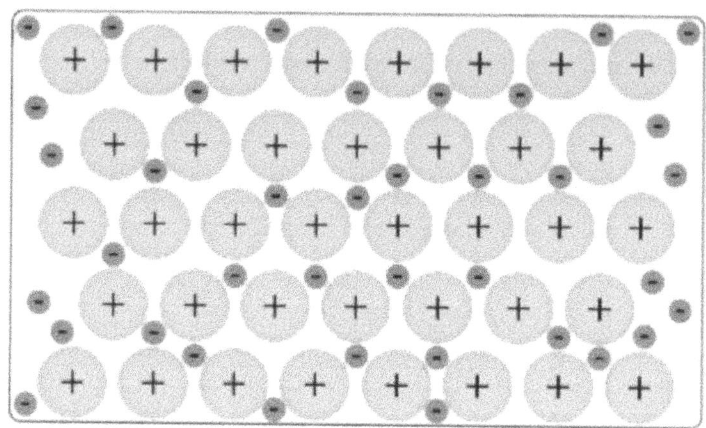

Metallic bonding. Positive ions (large circles) are held in regular patterns surrounded by valence electrons (dots) that move about freely within the metal.

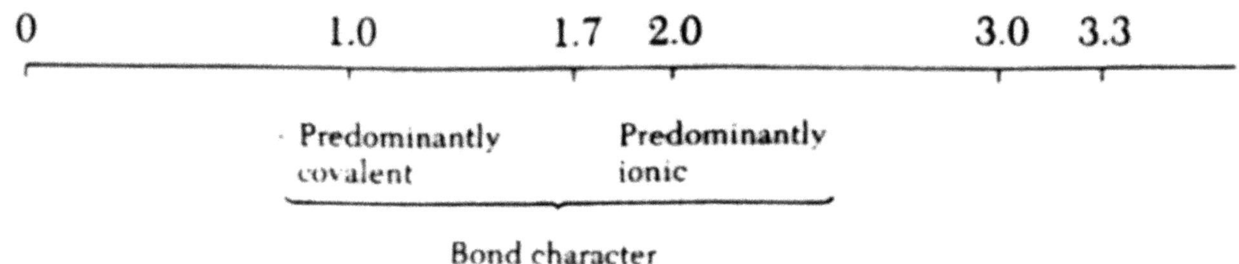

Electronegativity Differences

A chemist can use the electronegativity chart to predict bonding.

Refer to the electronegativity chart.

Cl_2 electronegativity for chlorine 2.0 – 2.0 = 0.00 or covalent bond
NaCl Cl = 3.0 – Na = 0.90 is equal to 2.0 or ionic bond

Ionization energies and electronegativies

Ionization Energies and Electronegativities

Top number: First Ionization Energy (kcal/mole of atoms); bottom number: Electronegativity

IA	IIA	IIIA	IVA	VA	VIA	VIIA	0
H 313 / 2.1							He 567
Li 124 / 1.0	Be 215 / 1.5	B 191 / 2.0	C 260 / 2.5	N 336 / 3.0	O 314 / 3.5	F 402 / 4.0	Ne 497
Na 119 / 0.9	Mg 176 / 1.2	Al 138 / 1.5	Si 188 / 1.8	P 254 / 2.1	S 239 / 2.5	Cl 300 / 3.0	Ar 363
K 100 / 0.8	Ca 141 / 1.0	Ga 138 / 1.6	Ge 187 / 1.8	As 231 / 2.0	Se 225 / 2.4	Br 273 / 2.8	Kr 323
Rb 96 / 0.8	Sr 131 / 1.0	In 133 / 1.7	Sn 169 / 1.8	Sb 199 / 1.9	Te 208 / 2.1	I 241 / 2.5	Xe 280
Cs 90 / 0.7	Ba 120 / 0.9	Tl 141 / 1.8	Pb 171 / 1.8	Bi 185 / 1.9	Po 2.0	At 2.2	Rn 248
Fr 0.7	Ra						

Cations and Anions (Salts and Ionic Bonding)

Cations are usually metals that lose electrons.

Li ion = $1s^2 2s^1$ Li⁺ cation = $1s^2$

Anions generally gain electrons and are often gases

F ion = $1s^2 2s^2 sp^5$ F- anion = $1s^2 2s^2 2p^6$

The above bonding forms salts because of the exchange of + and – charges. This is also ionic bonding.

Furthermore in ionic compounds the total positive charges of the cations must equal the total negative charges of the anions. The ionic compound for the salt NaCl (sodium chloride) shows that one formula unit of NaCl has one sodium ion and one chloride ion.

$$Na^+ + Cl^- \rightarrow NaCl$$

Library Topic

Write a report on salts (NaCl, KCl, NaF)

Some characteristics properties of acids and bases.

Acids	Bases
Electrolyte in water	Electrolyte in water
Causes certain indicators or dyes to change colors	Certain dyes will change colors
Tasteless sour (lemons)	Tastes bitter (quinine, peach seeds)
Proton donors (Bronsted-Lowry) H^+	Proton acceptors (Bronsted-Lowry) OH^-
Turns litmus paper red	Turns litmus paper blue
pH 0—6.9	pH 7.1—14
Loses hydrogen when it reacts with certain metals (Mg, Zn)	Feels slippery

Some common acids and bases.

Liquid	**Acid**	**pH**	**pH meter paper color**
Stomach acid	X	2.0	Red
Lemon juice	X	2.3	Red
Soda	X	3.0	Red
Citrus fruits	X	3.0—3.5	Red
Rainwater	X	6.2	Yellow
Milk	X	6.5	Yellow
Pure water	Neutral	7.0 (neutral)	Green
Blood	Base	7.4	Green
Seawater (concentrated)	Base	8.5	Blue
Milk of magnesia	Base	11.1	Blue
Drano or NaOH	Base	14	Dark Blue

LECTURE TOPICS

THE IDEAL GAS LAW

Lecture Topics

The three equations relating to pressure., volume and temperature can be combined into a single equation.

This equation is the ideal gas law: PV = nRT

where:

- P is pressure
- T is temperature
- R is 0.082056 L or $R = \frac{PV}{T} = \frac{1\,atm\,x\,22\,vl}{273°K}$
- N = moles

Some Uses of The Ideal Gas Law

Pressure: The ideal gas law can be used to find pressure.

Example A

Freon CF_2Cl_2 is used in air conditioners and refrigerators. If 8g of CF_2Cl_2 freon is injected into an empty 400ml container at 15°C. Find the pressure in atmospheres.

Solution $P = \frac{nRT}{V}$

$v = 400ml\,x\frac{1ml}{1000ml} = \frac{400ml}{1000ml}x\,liter = 0.4\,liter$

$T = 15°C + 273°K = 288°K$

$molar\,mass\,of\,CF_2Cl_2 = 8g\,x\,\frac{1mol\,CF_2Cl_2}{121g\,CF_2Cl_2} = 0.0661\,mole$

substitution: $P = \frac{nRT}{V} = \frac{0.0661mol\,x\,0.821l\frac{atm}{ml}x288°K}{0.4l} = 3.9atm$

FIND MOLES USING IDEAL GAS LAW

Note: ideal gas law = PV = nRT
to find moles = $n = \frac{P \, x \, V}{R \, x \, T}$

Example:

A travelling carnival has a large helium balloon. The technician records the initial volume at 60l. fills the balloon with He to a final pressure of 300 atm at 37°C. How many moles of He gas does the balloon contain?

Note: P = 300 atm V = 60l
T = 310°K (37°C = 273°K) R = 0.0821 atm mol/l

Solution $n = \frac{P \, x \, V}{R \, x \, T} = \frac{300atm \; x \; 60l}{0.0821 \text{ atm } x \; 310°K} = 707.24 \; moles$

Another Example of Moles

Canaries are beautiful birds, avid singers and sensitive to methane gas. Many coal miner count on canaries to indicate the poisonous gas methane CH4. A sick canary might indicate the presence of CH4.

Problem An underground cavern contains 1.64 x 10^5 liters of methane gas. The pressure is 18atm and the temperature is 32°C. How many moles of CH4 are present and how many grams of CH4.

Given V= 1.64 x 10^5 P = 18atm
R = 0.0211 atm K mol. T = 32 + 273 = 305°K

For moles $n = \frac{P \, x \, V}{R \, x \, T} = \frac{18atm \; x \; 1.64 \; x \; 10^5 l}{0.0821 \frac{L \, x \, atm}{k \, x \, mol} x \; 305°K} = 1.18 x 10^5 \; moles$

For grams $1.18 \; x \; 10^5 \; moles \; CH_4 \; x \; \frac{16g \; CH_4}{1 \; mol \; CH_4} = 1.88 \; x \; 10^6 \; grams \; of \; CH_4$

PROOF $\frac{1.18 \; x \; 10^5 \; grams}{16 \; g \; or \; 1 \; mol \; CH_4} = 1.16 \; x \; 10^5 \; moles$

PROOF $\frac{1.18 \; x \; 10^6 \; grams}{16 \; g \; or \; 1 \; mol \; CH_4} = 1.16 \; x \; 10^5 \; moles$

Another Example

The English system has an absolute scale which is called 459°R. R = Rankine pr 459°R. Conversion 15°F = 474°R, -10°F = 449°R.

Problem A given mass of Cl2 occupies 36 ft³ at 38°F. At constant pressure, find the new volume at 175°F.

Solution $V_2 = V_1 \; x \; \frac{T_2}{T_1} = 36ft^3 \; x \; \frac{634°R}{497°R} = 46ft$

Finding T_2 or New Temperature

Example 1

A chemist compresses H_2 in a cylinder. He has an initial volume of 50l and 80atm at 17°C. he cools the cylinder until the pressure is 12atm. Solve for the new temperature of H_2 in the cylinder. Volume remains at 50l.

Solution °K = 17°C + 273°K = 290°K

T = 290°K P1 = 80atm P2 = 12atm V1 and V2 = 50l

Solve for T2 $\frac{P_1 x V_1}{T_1} = \frac{P_2 x V_2}{T_2} = \frac{P_2 x T_1}{P_1} = \frac{12atm\ x\ 290°K}{80atm} = 43.5°K$

Convert to Celsius °L = K – 273

= 43.5 – 27_3

= -229.5°C

Example 2

A chemist compresses O_2 in a small cylinder. His initial volume is 30l and 20 atmospheres at 27°C. He warms the cylinder until the pressure is 30atm. Solve for the new temperature of O_2 in the cylinder. Volume remains constant.

Solution $\frac{P_2 x T_1}{P_1} = \frac{30\cancel{atm}\ x\ 300°K}{20\cancel{atm}} = 450°K$

°K to °C = 450°K – 273°K = 177°C

<u>ENGLISH SYSTEM</u>

14.7 lb/in^2 = 1 atmosphere.

<u>Example 1</u>

A pressure of 260 lb/in^2 was recorder using a pressure gauge of a hydrogen tank. Solve for atmospheres.

Solution $\frac{P_1}{P_2}\ x\ 1atm = \frac{28\ lb/in^2}{12\ lb/in^2} = 2.33 + 1atm = 3.33times$

Summary of Ideal Gas Law Lecture Topic

1. Provides a way to calculate moles or N of gases, temperature and volume.
2. Equation PV = nRT or P x V = n x R x T
3. Combined $\frac{P_1 x V_1}{T_1} = \frac{P_2 x V_2}{T_2}$

Lecture Problems for Ideal Gas Laws

1. A chemists inject 19g of CF_2Cl_2 into an empty 600ml container at 49°C. What is the new pressure in atmospheres?
2. A student records a plastic bag volume of 30l. he fills the bag with N_2 to final pressure of 150atm at 22°C. how many moles of N_2 gas does the bag contain?
3. A cave contains 2.13 x 104 liters of SO_2. The pressure is 12atm and the temperature is 21°C. How many moles of SO_2 are present?
4. A research chemist compresses CO_2 in a glass cylinder. She has an initial volume of 30l and 56atm at 19°C. She cools the cylinder until the pressure is 14atm. What is the new temperature of CO_2 if volume is constant?
5. Prediction
   Explosion will occur between 3.78atm and 4.1atm.

   A terrorist finds an aerosol can. The pressure is 1atm and 30°C. He throws the can into a fire. The final temperature of the fire is 888°C. Find the pressure. Will the can explode?

6. A volume of $H_{2(g)}$ occupies 16ft^3 at 14°F. At constant pressure find the new volume 78°F.
7. A new car tire contains air at 42lbs per in^2. How many times the original volumes would the air occupy if released at 17lb per in^2?

Note: tire gauge registers one extra atmosphere.

8. Convert: 76 cm to atmosphere
   5.6l to moles
   ___°R 25°F

LECTURE TOPICS GAS LAWS: PART 2

Chemist can use the ideal gas law to calculate the molecular weight of gases.

The ideal Gas Law is PV =nRT

P = pressure V = volume n = moles R = 0.0821atm

Sample Problem I Finding Molecular Weight

3.75g of a gas occupy a volume of 2.15l at a temperature of 70°C and a pressure of 1.35 atmospheres. Find the molecular weight.

$$M = \frac{m \, x \, R \, x \, T}{P \, x \, V}$$

$$\text{Molecular Weight} = \frac{3.75g}{1.35atm} \, x \, 0.0821 \, x \, \frac{1atm \, x \, 343°K}{mole \, °K \, x \, 2.15 \, liters} = 36.4g$$

For Further Study

Reasoning Method

Same chemists use logical reasoning to find the molecular weight of a gas. The previous problem can also be worked as follows.

Previous Problem Restated

3.75 grams of a gas occupy a volume 2.15l after temperature is decreased from 70°C to 0°C and pressure is decreased from 1.35atm to 1atm. Find the M.W. weight of the gas.

Solution:

1. First Solve for Volume

$$V_2 = V_1\, x\, \frac{T_2}{T_1}\, x\, \frac{P_1}{P_2}$$

$$V_2 = 2.15l\, x\, \frac{273°K}{343°K}\, x\, \frac{1.35atm}{1.00atm} = 2.32liters$$

2. Convert liters to grams or molecular weight.

$$1\, mole\ =\ 22.4l\ STP\ x\ \frac{3.75g}{2.32l} = 36.4grams$$

The Search For Volume

A diver filled a small tank with O_2. She did not record the volume. However she noted 2.11g of O_2, 4.0 atmospheres and 18°C. Solve for volume.

Solution

$$\frac{2.11\ of\ O_2}{32g} = 0.66\ moles \qquad °K = 18 + 273 = 291°K$$

$$V = \frac{nRT}{P} = \frac{0.66\ x\ 0.821\frac{mol}{l}\ x\ 291°K}{4atm} = \frac{1.58}{4} = 0.394\ literr\ or\ 394ml$$

PROOF

A proof can easily be calculated. One can solve for pressure.

$$P = \frac{nRT}{V} = \frac{0.661\ x\ 0.0821l\ x\ 291°K}{0.394\ liter} = 4atm$$

GRAHAM'S LAW OF DIFFUSION

A simple method of finding an unknown gas.

Review:

According to Graham the lighter the gas the faster they diffuse. On other hand the heavier the gas the slower they diffuse. For more a Graham's of Diffusion.

What are the relative rates of diffusion of $Cl_2(g)$ and $H_2(g)$. Write heavier over lighter gas and get square root.

$$\frac{\sqrt{71}\,Cl_2}{2H_2} = \sqrt{35.5}\,grams = 5.96$$

H_2 is 5.96:1 faster than $Cl_2(g)$

Sometimes a chemist is required to find unknown gases. One could use a simple formula to find the weight of an unknown gas and then the formula.

Formula for heavier gas: $rate^2$ x weight of lighter gas

Example A: Relative rate of H_2 is 5.96:1 faster than gas X. Find gas X.

Solution $rate^2$ x weight of lighter gas
$= (5.96)^2 = 35.5 \times 2 = 71g$ or Cl_2 (g)

Example B: A technician recorded a relative rate of diffusion of 2:1. He remembered using CH_4 but forgot to record if he used SO_3 H_2S orSO_2. Find the gas he used.

Solution $rate^2$ x weight of lighter gas
22 = 4 x wt of CH_4 = 4 x 16g =64 grams
The chemist used SO2 (sulfur dioxide) because wt of SO_2 is 64 grams.

FINDING LIGHTER GASES

To find the weight of the lighter gas square the rate of the lighter gas and divide.

Example Cl_2 and H2 $\frac{\sqrt{71}}{2} = 35.5$

$\sqrt{35.5} = 5.96{:}\,1$

$\frac{71g}{35.52g} = 2g$

For lighter gas $(5.96)^2$

Lecture Topics

Gas Laws Part 2 Problems

1. What are the relative rates of diffusion of CH4 $_a$nd H_2 gases?
2. A student records a relative rate of diffusion of 1.6:1. The lighter gas is NH_3(g). Find the weight of the heavier gas.
3. A small cylinder is filled with H_2. The technician failed to record the volume. He did remember 5.2g of H_2, 3 atmospheres and 28°C. What was the volume.
4. A college professor noted that 1.75 grams of a gas occupy a volume of 7 liters at a temperature of 38°C and a pressure of 1.80 atmosphere. Find the mole molecular weight of the gas (note: 38°C is T and 1.80atm is P).

Lecture Topics 2

A. Light as Energy, Particles and Waves

Max Planck

A German physicist observed the light and heat given off by a hot body. He explained his observation by saying that light is made of discrete bundles of energy. Planck called these bundles quanti (plural of quantum).

Moreover the amount of energy in each quantum depends on the color of light. For example blue light has a greater quantum of energy than red light. Quantum or photons are the fundamental units of lights.

The frequency and wavelength of light waves are inversely related. As the wavelength increases, the frequency decreases. The wavelength and frequency do not affect the amplitude.

Long wavelength means low frequency.
Short wavelength means high frequency.

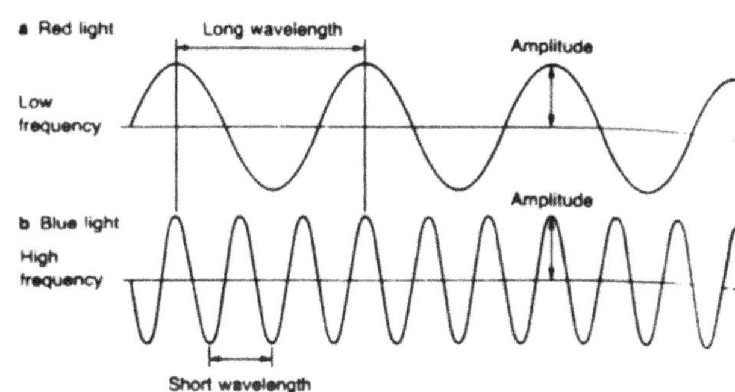

Key Terms

- Wavelength – the distance between neighboring, peaks or thoughts
- Frequency – the number of peaks that pass a given point each second.
- Wave velocity – the distance a peak moves in a unit of time (usually) 1 second

The electromagnetic spectrum consists of radiation over a broad band of wavelengths. The visible light portion is very small. It is in the 10-7m wavelength range and 1015 hertz (s-1) frequency range. What types of nonvisible radiation have wavelengths close to those of red light? To those of blue light?

Infrared is close to red. Ultraviolet is close to blue

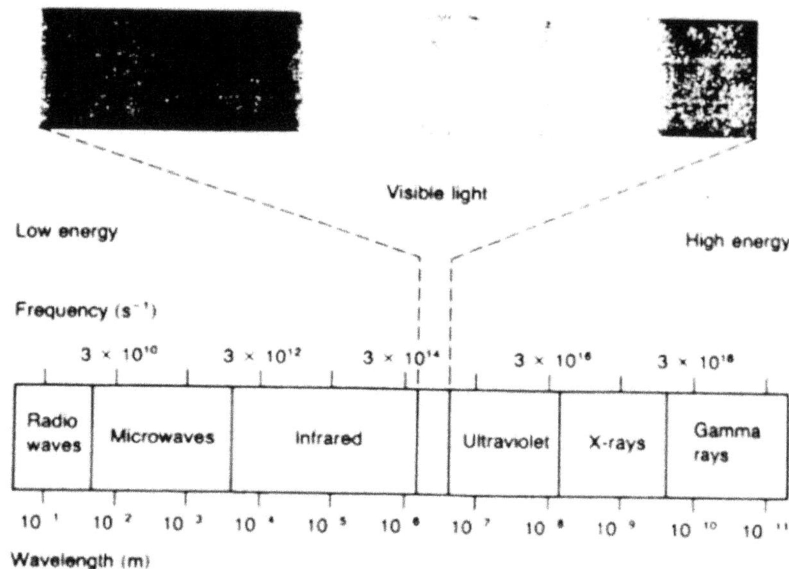

Some Common Symbols

- V = velocity
- F = frequency (Hertz)
- λ = wavelength
- V = F λ

Wavelength Problems

Sample 1 Velocity Problem

What is the velocity of a wave with a frequency 14 Herts and a wavelength of 7.0 meters?

Solution V = F λ

= 14 waves/second x 7.0 meters/wave

= 98 meters/second

= 9.8×10^1 m/s

Sample 2 Velocity Problems

What is the velocity of a wave with a frequency of 19 Hertz and a wavelength of 4.3 meters?

Solution V = F λ

= 19 waves/second x 4.3 meters/waves

= 81.7 meters/second

= 8.17×10^1 m/s

FREQUENCY PROBLEMS

Scientists can measure the wavelengths of light in a bright line spectrum. This is because each color has a particular frequency. A scientist can calculate the frequency if he knows the wavelength and uses the formula c = f λ.

Note c is the speed of light, F = frequency

The speed of light or c in a vacuum tube is 3.0 x 10^8 meters/second pr 3.0 x 10^{10} cm/second

Same scientist calculate the speed of light as 186,000 miles per second in a vacuum tube.

Note: Simplified conversion

(5280 ft = 1 mile, 39 inches = 1 meter) for speed of light in a vacuum tube.

$$\frac{5280\ ft}{1\ miles}\ x\ \frac{12\ inches}{1ft} = \frac{63360\ inches}{39\ inches}\ x\ \frac{186{,}000\ m/sec}{1\ mile/sec}\ x\ 1\frac{meter}{sec} = 3.02\ x\ 10^8\frac{meter}{second}$$

Sample B Frequency of Light Problem

Calculate the frequency of a quantum of light (or photon) with a wavelength of 5.0 x 10^{-6} meters.

Solution $$F = \frac{c}{\lambda} = \frac{3.0\ x\ 10^8\ m/s}{3.0\ x\ 10^8\ m/s} = F\ x\ 5.0\ x\ 10^{-6}\ meters$$

$$F = \frac{3.0\ x\ 10^8\ m/s}{5\ x\ 10^{-6}\ m/wave}$$

$$F = 0.6\ x\ 10^{14}$$

$$F = 6.0\ x\ 10^{13}\ Hertz$$

PROOF $$\text{wavelength or } \lambda = \frac{C}{F} = \frac{3.0\ x\ 10^8\ m/s}{6.0\ x\ 10^{-13}\ Hz}\ x\ 1\ meter = 5.0\ x\ 10^{-6}\ meter$$

Calculate the wavelength of a quantum of light with a hertz of 4.0 x 10^{13}

Solution $$\text{wavelength or } \lambda = \frac{C}{F} = \frac{3.0\ x\ 10^8\ m/s}{4.0\ x\ 10^{13}\ m/s} = 7.5\ x\ 10^{-6}\ meters$$

PROOF $$F = \frac{C}{\lambda} = \frac{3.0\ x\ 10^8\ m/s}{7.5\ x\ 10^{-6}\ m/s} = 7.5\ x\ 10^{13}\ Hertz$$

Energy of Light Problem

The energy of light is directly proportional to the frequency of the radiation. Scientist use a constant of proportionality. Called Planck's constant of h. The formula for energy is:

$$E = hf$$

Note: E = energy in joules h or Planck's constant is

$$h = 6.6 \times 10^{-34} \text{ joule/hertz}$$

Energy of Light Problem

Calculate the energy of light or photons. if radiation has a frequency of 4.0×10^{12} hertz

Solution Used $E = h \times f$
$= 6.6 \times 10^{-34}$ Joule/Hertz $\times 4.0 \times 10^{12}$ Hz
$= 2.64 \times 10^{-21}$ joule/hertz

CONSUMER AND ENERGY

Millions of people benefit from light as energy. particles and waves. Records, neon lights, cassettes, television and movies are just a few example of light energy. Other examples are lasers, solar calculators, solar radios and solar heated houses.

Some uses of photocells by Einstein

Problems and Questions

1. Use a reference book to describe some contribution of Einstein, Max Planck, Fermi, Millikan, Joule, Marconi and DeForest.

2. Define the following:
   - photons
   - troughs
   - frequency
   - wave velocity
   - quantum

3. Calculate the Frequency of a quantum of light (or photon) with a wavelength of 7.0×10^{-5} meters.

4. A scientist accidentally deleted his light Frequency on his super computer. He did record a wavelength of 2.75×10^{-4} meters. Find the deleted light Frequency.

5. Calculate the energy of light or photons if the Radiation Frequency is 5.8×10^{13} Hertz.

6. What is the velocity of a wave with a frequency of 9 Hertz and wavelength of 4.27 meters.

A Few Physical Properties of Solution Ideal Solution

Chemists can use various physical properties analyze and identify solutions. Some basic techniques are boiling points, freezing points, density, and vapor pressure. A solution that gives true physical property results in an *ideal solution*. Pure H_2O boils at 100°C and melt at 0°C.

Moreover in reality there are few ideal solutions. But many dilute solutions are almost ideal. *The less solute present the closer the solvent is to ideal or exact physical properties.*

Raoult's Law:

According to Raoult's Law the vapor pressure of a volatile solvent in a dilute solution is proportional to its mole fraction.

Simplified

A chemist adds 3 moles of sucrose to 60 moles of H_2O. the vapor pressure of H_2O will be at 31.51mmHg at 30°C.

V.P. sol. x 31.51mmHg 30mmHg

$$= \frac{60\ \cancel{moles}\ of\ H_2O}{63\ \cancel{moles}\ of\ sub + H_2O}$$

$$V.P.sol. = \frac{60}{30}\ x\ 31.51\ mmHg = 30mmHg$$

Molecular Weight of Solute

Some substance are not capable of being vaporized. We can use Raoult's law to find the molecular weight of a solute.

Example The addition of 26g of a solute to 125g of H_2O lowers the V.P. at 25°C from 23.52 torr to 21 torr. Find the molecular weight of the solute.

Step 1 Mole fraction H_2O $\frac{VP\ Solution}{VP\ H_2O} = \frac{21\ torr}{23.52} = 0.893\ mole\ H_2O$

Step 2 Moles Fraction Solute = 1.0 – 0.893 = 0.107

Moles Fraction Solute 1.0 0.893 0.107

Moles Solute: $\frac{moles\ H_2O}{0.893} - moles\ H_2O = \frac{0.107}{0.893}\ x\ moles\ H_2O$

$$= \frac{0.107}{0.895} x\ 125g\ moles\ H_2O\ x\ \frac{1\ mole\ moles\ H_2O}{18g\ moles\ H_2O} = 0.83$$

$$molecular\ weight = \frac{26g}{0.83\ moles} = 31.3g$$

VAPOR PRESSURE PROBLEM

Example 1 Two moles of sucrose ($C_{12}H_2O_{11}$) is added to 22 moles of H_2O at 30°C. The vapor pressure of H_2O at 30°C is 31.51 mmHg. Find the vapor pressure of this mixture.

Solution:

$$V.P. = \frac{Number\ of\ moles\ A}{Total\ number\ of\ moles}$$

$$= \frac{22\ moles\ of\ H_2O}{24\ moles\ of\ substance\ (sugar + H_2O)}$$

$$V.P.sol.n. = \frac{22}{24}\ x\ 31.51mmHg = 28.85mmHg$$

Molecular Weight of Solute Osmotic Pressure

Osmosis is defined as movement of molecules from greater to lesser concentration through a membrane in some cases molecules cannot pass through the walls of a membrane. This blockage will cause the solution to effect a pressure against the walls. Chemist call this osmotic pressure. We can use osmotic pressure to find weight.

Symbols: Π = osmotic pressure
n = moles
v = volume
R = 0.0821
T = °C + 273°K
c = concentration

Formula for osmotic pressure is $c = \frac{\Pi}{RT}$

Example: A solution has 60g of protein. This solution has an osmotic pressure of 7.3 torr at 25°C. Find the molecular weight of the protein.

Solution $c = \frac{\Pi}{RT}$

Substituting $\frac{7.3\ torr\ x \frac{1\,atm}{760\ torr}}{0.0821\ x \frac{1atm}{mole°K} x\ 298°K} = \frac{7.3torr}{760\ x\ 0.821\ x\ 298} = 3.926\ x 10^{-4}\ mole/l$

$$molecular\ weight = \frac{grams}{moles} = \frac{60g}{3.93\ x\ 10^{-4}} = 1.53\ x\ 10^{5} \frac{g}{mole} = \frac{153.000g}{moles}$$

Physical Properties of Solutions Problems

1. Define:
    - solvent
    - solute
    - osmosis
    - manometer
2. The vapor pressure of H_2O at 30°C is 31.51 torr. Find vapor pressure (V.P.) if 7 moles of glucose is added to 35 moles of H_2O.
3. A chemists adds 43 grams of $C_6H_{12}O_6$ to 312g of H_2O. Find the vapor pressure of the solution.
4. A student noticed the addition of 42g of a solute to 140g of H_2O lowered the vapor pressure at room temperature from 23.52 torr to 19 torr. Find the molecular weight of the solute.
5. A solution contain 50g of protein per liter exerts an osmotic pressure of 8.2 torr at 25°C. Find the molecular weight of this protein.

Colligative Properties for Non H_2O Solvents

Review

We studied the boiling points and freezing points for pure H_2O and nonelectrolyte solvents. For pure H_2O b.p. 1s 100°C and 1ts F.p. 0°C. However a solute such as sugar will raised the b.p. and lower the F.p. of H_2O.

Example

1 mole of sucrose will raise the b.p. of H_2O to 100°C or 100.53°C. Likewise 0.5 moles of sucrose $C_{12}H_{22}O_{11}$ will lower the b.p. to 100.26°C or 0.5 moles x 0.52 + 100°C - 100.26°C.

A. 0.6 molal solution of sucrose in H_2O should freeze at

0.6 molal x -1.86 K.f.p. = -1116°C
Sample of Non H_2O solvents

Benzene C_6H_6

A college student dissolves 65g of an unknown compound in 230g of benzene. The b.p. is raised by 1.48°C.

Find the molecular weight of this compound.

K.B.P. of C_6H_6 is 2.53°C Kg/mole solute.

Solution: The b.p.is raised by 1.48°C or $\frac{1.48°C}{2.53°C}$ or 0.585 mole/solute

Step 2 Mass 1Kg b.p
g of solvent moles/solute

$$65\ g\ x\ \frac{1000g}{230}\ x\ \frac{2.53°C}{1.48°C} = 482.7\ grams/mole$$

A CARBON TETRACHLORIDE $CC1_4$

A technician dissolves 52g of x in 550g of CC1. The boiling point increases by 7.10°C. Find the weight of compound x. K.B.P. of $CC1_4$ =5.03°C mole/solute

Solution: mass 1Kg / g of solvent K b.p. / b.p./mole/solute

$$= 52g\ x\ \frac{1000g}{550g}\ x\ \frac{5.03°C}{7.10} = 65.53\ grams/mole$$

<u>Freezing Points</u>

What is the molecular weight of unknown compound x if 12g are dissolved in 75g of benzene. The solution F.p. is depressal by 8.4°C. Note K.F.p. of C6H6 is 4.9°C - Kg.

$$12\ x\ \frac{1000g}{75g}\ x\ \frac{4.9°C}{8.4°C} = \frac{93g}{mole}\ solute$$

Note General Formula

Mass of solute 1 Kg / grams of solvent K(b.p) or K (F.p.) / b.p elevation or F.p. depression

<u>Problems</u>

1. A solution has 19 grams of a nonelectrolyte in 220g of H_2O. The solution boils at 100.13°C. Find the weight of the solute. What is the freezing point of this solution?
2. When 14g of an unknown are dissolved in 45g of benzene (C_6H_6) the F.p. is depressed by 6.8°C. Calculate the weight of the solute. Note K.F.p. of C_6H_6 is 4.9°C - Kg
3. A student dissolve 16g an unknown is compound in 165g of $CC1_4$. The b.p. changes to 9.3°C. Find the molecular weight of this compound. K.B.p. of ($CC1_4$) -5.03°C - Kg $CC1_4$/mole solute.

Rates of Chemical Reactions

The rate of chemical reaction refers to the amount of matter used or produced per unit of time.

For example if the total volume is tripled than the amount of matter produced per second is tripled provided other conditions are constant.

Definition: Specific reaction rate is equal to the amount of matter undergoing changes per second or per volume of the reaction system.

The unit for specific reaction rate is moles per liter second.

Example: $H_2 + C1_2 \rightarrow 2HCl$

In a 35 liter vessel at 290°C the rate of HCl. production is 5.0 x 10^{-3} mole HCl sec.

Find the reaction rate S.R.R. $\frac{Rate\ of\ Reaction}{Volume}$

$$= \frac{5.0\ x\ 10^{-3}\ mole\ HCl/sec}{35\ liters} = 1.43\ x\ 10^{-4}\ mole \frac{HCl}{liters} sec$$

Law of Mass Action or Rate Dependence on Contraction

Many chemists believe it is difficult to determine if the rate of reaction is based on reactants or products. This is because reactions rates vary based on temperature, concentration and pressure. Moreover despite these variables it is found that concentration influences most reaction rate. The coefficients in the equations are raised to exponents.

Example of Law of Mass Action

$H_2 + Cl \neq 2\ HCl$

Specific Rate (production of HCl) = Kf x CH_2 x CCL_2

Reverse Reaction for $H_2 + Cl_2 \rightarrow 2HCl$ = specific rate (decomposition of HCl) = Kf x C_2 HCl

Note Kf and Kr are specific rate constants.

Kf = forward reaction

Kr = reverse reaction

<u>Example</u> If the specific rate constant for HI at 400°C is .43 find the specific rate of formation of HI at 400°C if.

$$I_2 = 0.008\frac{mole}{l} and\ H_2 = 0.02mole/l$$

Formula $S.R.R. = K \times CH_2 \times CI_2$
$= 0.43 \times 0.2 \times 0.008$
$= 6.88 \times 10^{-5}$ mole HI/l sec

Note: We should emphasize that same reaction rates do not follow the law of mass action. Therefore a technician must first determine if the law of mass action is applicable.

Mechanisms of Chemical Reactions

One might conclude that if reaction rate is based on concentration then reaction Occurs by collision of molecules. In other words collision of molecules is directly proportional to concentration. This is not true.

Moreover what occurs is a rapid reversible equilibrium formation of ions or atoms.

Example: The formation of Br atoms from Br_{2}.

1. $Br_2 \neq 2\ Br$ (very fast reaction)
   Note Br unities with H_2
2. This form 2HBr: or $2Br\ H_2 \neq [Br..H..Br..H] \rightarrow 2HBr$

Another Reaction Example

$N_2O_5 \neq NO_3 + NO_2$ slow
$NO_3 + NO_{2 \rightarrow} NO_2 + NO_3$ slow
$NO_3 \neq NO + O_2$ fast
$NO + NO_3 \rightarrow 2NO_2$ fast
$2NO_2 \neq NO_2O$ fast

Note despite five steps in the decomposition of $N_2O_5 \rightarrow N_2O_4$ the reaction rate is determined by the slow step or N_2O_5

Question and Problems

1. Describe the law of mass action.
2. Find the specific reaction rate for the following. 65 liter vessel at 280°C has a rate of production of HBr at 7.0 x 10^{-3} mole HBr/sec.

$$\mathbf{Br_{2(g)} + H_{2(g)} \rightarrow 2HBr_{(g)}}$$

3. A chemist's has a new compound. The formula is XY. What is the specific rate of formation of XY in a vessel at 300 C. Note compound XY has a rate constant formation of 0.26 (mol/Liters)

$X_2(g)$ = 0.05 mole/liter
$X_2\ (g) + Y_2\ (g) \rightarrow 2XY(g)$
$Y_2\ (g)$ = 0.008 mole/liter

Easy Chem
Quiz Metric System

Name: ____________________

A. Convert the Following:

150°C ______°F

60°C ______°F

485°F ______°C ______°A

1.75 liters ______ ml

375°A ______°C

195°C ______°F

160mm ______ cm ______m

30 decimeters ______ m

B. What is the density of X if a technician has 160cc of H_2O and 30 grams of X?

C. A chemist's has 135cc of element x and a density of 16g/cc.

D. How many cc does a sophomore need to prepare 60 grams of x if the density is 0.4 g/cc.

EASY CHEM QUIZ

Name: ____________________

Percent Composition

1. A Ring consist of 60% Au. 17% Ag, 10% A1 and 13% Cu.

Solve The Following

<u>Ring Weights 28 grams</u>

1. Grams of Au ______, grams of Ag ______, grams of Al ______, grams of Cu ______.
2. A jeweler has a beautiful stone. But his analysis is incomplete. The gem weight 43 grams. He has 10.75 grams of Au or ______ %, 12.47 grams of Cu or ______%. He also found ______ grams of Ag or ______ %.
3. Find the C in glucose $C_6H_{12}O_6$.
4. Find the % of S in sulfuric acid H_2SO_4.

Gas Laws Quiz

1. What is the density of laughing gas N_2O at S.T.P.
2. Calculate a student has 11.2 liters of gas *x* with a density of 7.5 g/liters. Find the mass of the gas (note 11.2 liters in 5 moles of a gas).
3. What is the new volume of 600 ml O_2 if you increase the pressure from 760 torr to 900 torr, temperature constant.
4. Find new volume 250 cc of O_2 is collected over water at 19°C to 30°C and 750 torr to 780 torr. Find the new volume.
5. An analytical chemists has 800cc of gas *x* who has H, C, N. She has 60% Hydrogen ______ or cc, 130cc of Nitrogen a ______ %, C is 24% or ______ cc.
6. A technician has 125 ml of halothane gas. He changes the pressure 780 to STP and temperature 15°C to 75°C. Find the new volume of halothane?
7. Complete the following chart based on H_2S gas.

Moles	Liters	Molecules	Grams
A.			14
B. 0.75			
C.	13.44		
D.	44.8		

8. A diver spots a shark and climbs to the surface so fast or excess N, in his blood. Analysis showed he inhaled 135 for 6 seconds. How many ml of N_2 ______, moles of N_2 ______, and molecules of N2 ______ did he use?
9. What are the relative rates of diffusion of SO_2 and H_2S?
10. S+ O2 → SO_2
    a. to increase S you must ______ pressure
    b. to decrease SO_2 you must ______ pressure.

Thermochemistry Quiz

Name: ____________________

1. How many calories and joules are needed to convert 70 grams of H_2O from 50°C to 90°C.
2. The specific heat of metal x is 0.165 cal/g°C. How many joules (and calories) are required to heat 35 grams of the metal from 12°C to 85°C?
3. A student has a block of ice. She used 2400 calories to melt the ice from 0°C liquid. How many grams of ice did she have? How many moles of ice?
4. How many calories (also joules) are needed to convert 1.8 moles of ice (wt. of H_2O) from -60°C to 115°C?
5. A college student dropped 40 grams of element x at 35°C in 75 grams of H_2O at 55°C. The final H_2O temperature is 66°C. What is the specific heat of this metal?

Easy Chemistry Quiz

Name: ____________________

Solutions

1. Find molarity if 65 grams of $C_6H_{12}O_6$ are dissolved in 125 cc of H_20.
2. How many grams of $C_6H_{12}O_6$ are needed to prepare 400 mL of a 0.2m solution.
3. A chemists records a b.p. of 100.26°C (electrolyte). He added 60 grams of the compound to 500 grams of H_2O. Find molarity and weight. What is the freezing point of this solution?
4. An electrolyte freezes at -3.72°C. The compound is NaCl = Na^+ + $C1^-$. What is the molarity of this compound? What is the boiling point?
5. A physician discovers a new antibiotic. One mole weights 40 grams. Complete the following.

   A. 5 x 10^{23} molecules ________ grams
   B. 4.5 moles in 80 ml ________ grams ________ molarity
   C. .5 moles in 200 ml = ________ molarity

THE MATHEMATICS OF CHEMICAL EQUATIONS

Find the answer to each problem and write it in the space at the right. In solving the problems, use the table of atomic masses below.

calcium, Ca	40.0 u	nitrogen, N	14.0
carbon. C	12.0	oxygen, O	16.0
chlorine. Cl	35.5	platinum, Pt	195
hydrogen, H	1.01	potassium, K	39.1
iron, Fe	55.8	sodium, Na	23.0
magnesium, Mg	24.3	sulfur, S	32.0

$$2H_2O + 2H_2 + 0_2$$

1. When water, H_2O is decomposed, it produces hydrogen gas. H_2, and oxygen gas, O_2. In order to produce 4.0 moles of oxygen gas, how many moles of water must be decomposed?
2. In the reaction $FeS + 2HC1 \rightarrow FeC1_2 + H_2S$. How many moles of HCI are required produce 5 moles of H_2S?
3. A 40g sample of a gas has a volume of 5.6l at STP. What is its molecular mass?
4. Ammonia burns in oxygen to produce nitric oxide and water, as represented by the equation $4NH_3\ (g) + 50_2(g) \rightarrow 4NO(g) + 6H_2O(g)$. If 10 L of ammonia are burned, what volume of steam is produced a constant temperature and pressure?
5. In the reaction, $Mg(s) + 2HCI \rightarrow MgCl_2 + H_2\ (g)$, how many liters at STP of hydrogen are produced by 48.6 grams of magnesium?
6. 36l of any gas ________ moles.
7. $3H_2\ (g) + N_2\ (g)\ 2NH_3\ (g)$ 6 liters of H_2 gas yield ________ moles of NH_3 gas.
8. $4P + 5O_2 \rightarrow 2P_2O_5$ 0.8 moles of P________ moles of $2P_2O_5$.

LECTURE QUIZ ON MOLES

Part 1

1. Convert to moles

   1.0×10^7 atoms of Ag ______ moles.

   Convert to grams

   1.0×10^7 atoms of Ag ______ grams.
2. Solve for the number of atoms in 100 mg of Al metal.
3. The U.S. debt could rise to 5.5 trillion dollars. Convert to pennies then to moles of pennies.
4. Convert 0.65g of $C_9H_8O_4$ aspirin to moles ______ molecules.
5. A chemist 410g of insecticide the molecular weight 65g. Three moles per acre are needed. How many acres can she cover ________?

Part 2 Weight of insecticide is 65g.

Convert 5g of insecticide to ______ moles ______molecules.

Part 3

4 atoms ______ moles ______ grams.

Element	Symbol	Atomic Number	Atomic Mass	Melting Point (°C)	Boiling Point	Density (g/cm3) (gases at STP)	Major oxidation states
Actinium	Ac	89	(227.0482)	1050	3200	10.07	+3
Aluminum	AI	13	26.98154	660.37	2467	2.6989	+3
Americium	Am	95	243	994	2607	13.67	+3, +4, +5, +6
Antimony	Sb	51	121.75	630.74	1950	6.691	-3, +3, +5
Argon	Ar	18	39.948	-189.2	-185.7	0.0017837	
Arsenic	As	33	74.9216	871	613	5.73	-3, +3, +5
Astatine	At	85	(210)	302	337	—	
Barium	B	56	137.33	725	1640	3.5	+2
Berkelium	Bk	97	(247)	986	——	14.78	
Beryllium	Be	4	9.01218	1278	2970	1.848	+2
Bismuth	Bi	83	208.9804	271.3	1560	9.747	+3, +5
Boron	B	5	10.81	2079	3675	2.34	+3
Bromine	Br	35	79.904	-7.2	58.78	3.12	-1, +1, +5
Cadmium	Cd	48	112.41	320.9	765	8.65	+2
Calcium	Ca	20	40.08	839	1484	1.55	+2
Californium	Cf	98	(251)	900	—	14	
Carbon	C	6	12.011	3550	4827	2.267	-4, +2, +4
Cerium	Ce	58	140.12	799	3426	6.657	+3, +4
Cesium	Cs	55	132.9054	28.40	669.3	1.873	+1
Chlorine	Cl	17	35.453	-100.98	-34.6	0.003214	-1, +1, +5, +7
Chromium	Cr	24	51.996	1857	2672	7.18	+2, +3, +6
Cobalt	Co	27	58.9332	1495	2870	8.9	+2, +3
Copper	Cu	29	63.546	1083.4	2567	8.96	+1, +2
Curium	Cm	96	(247)	1340		13.51	+3
Dysprosium	Dy	66	162.5	1412	2562	8.550	+3
Einsteinium	Es	99	(252)	—	—	—	
Erbium	Er	68	167.26	159	2863	9.066	+3
Europium	Eu	63	151.96	822	1597	5.243	+2, +3
Fermium	Fm	100	(257)	—	—	—	
Fluorine	F	9	18.998403	-219.62	-188.54	0.001696	-1
Francium	Fr	87	(223)	27	677		+1
Gadolinium	Gd	64	157.25	1313	3266	7.9004	+3

Gallium	Ga	31	69.72	29.78	2403	5.904	+3
Germanium	Ge	32	72.59	937.4	2830	5.323	+2, +4
Gold	Au	79	196.9665	1064.43	3080	19.3	+1, +3
Hafnium	Hf	72	178.49	2227	4602	13.31	+4
Helium	He	2	4.00260	-272.2	-268.934	0.001785	
Holmium	Ho	67	164.9304	1474	2695	8.795	+3
Hydrogen	H	1	1.00794	-259.14	-252.87	0.00008988	+1
Indium	In	49	114.82	156.61	2080	7.31	+1, +3
Iodine	I	53	126.9045	113.5	184.35	4.93	-1, +1, +5, +7
Iridium	Ir	77	192.22	2410	4130	22.42	+3, +4
Iron	Fe	26	55.847	1535	2750	7.874	+2, +3
Krypton	Kr	36	83.80	-156.6	-152.30	0.003733	
Lanthanum	La	57	138.9055	921	3457	6.145	+3
Lawrencium	Lr	103	(260)	—	—	—	+3
Lead	Pb	82	207.2	327.502	1740	11.35	+2, +4
Lithium	Li	3	6.941	180.54	1342	0.534	+1
Lutetium	Lu	71	174.967	1663	3395	9.840	+3
Magnesium	Mg	12	24.305	648.8	1090	1.738	+2
Manganese	Mn	25	54.9380	1244	1962	7.32	+2, +3, +4, +7
Mendelevium	Md	101	257	—	—	—	+2, +3
Mercury	Hg	80	200.59	-38.842	356.58	13.546	+1, +2
Molybedum	Mo	42	95.94	2617	4612	10.22	+6
Neodymium	Nd	60	144.24	1021	3068	6.9	+3
Neon	Ne	10	20.179	-248.67	-246.048	0.0008999	
Neptunium	Np	93	237.0482	640	3902	20.25	+3, +4, +5, +6
Nickel	Ni	28	58.69	1453	2732	8.902	+2, +3
Niobium	Nb	41	92.9064	2468	4742	8.57	+3, +5
Nitrogen	N	7	14.0067	-209.86	-195.8	0.0012506	-3, +3, +5
Nobelium	No	102	(259)	—	—	—	+2, +3
Osmium	Os	76	190.2	3045	5027	22.57	+3, +4
Oxygen	O	8	15.9994	-218.4	-182.962	0.001429	-2
Palladium	Pd	46	106.42	1554	2970	12.02	+2, +4
Phosphorus	P	15	30.97376	44.1	280	1.82	-3, +3, +5
Platinum	Pt	78	195.08	1772	3827	21.45	+2, +4

Plutonium	Pu	94	(244)	641	3232	19.84	+3, +4, +5, +6
Polonium	Po	84	(209)	254	962	9.32	+2, +4
Potassium	K	19	39.0982	63.25	760	0.862	+1
Praseodymium	Pr	59	140.9077	931	3512	6.64	+3
Promethium	Pm	61	(145)	1168	2460	7.22	+3
Protactinium	Pa	91	231.0359	1560	4027	15.37	+4, +5
Radium	Ra	88	226.0254	700	1140	5.5	+2
Radon	Rn	86	(222)	-71	-61.8	0.00973	
Rhenium	Re	75	186.207	3180	5627	21.02	+4, +6, +7
Rhodium	Rh	45	102.9055	1966	3272	12.41	+3
Rubidium	Rb	37	85.4678	38.89	686	1.532	+1
Ruthenium	Ru	44	101.07	2310	3900	12.41	+3
Samarium	Sm	62	150.36	1077	1791	7.520	+2, +3
Scandium	Sc	21	44.9559	1541	2831	2.989	+3
Selenium	Se	34	78.96	217	684.9	4.79	-2, +4, +6
Silicon	Si	14	28.0855	1410	2355	2.33	-4, +2, +4
Silver	Ag	47	107.8682	961.93	2212	10.50	+1
Sodium	Na	11	22.98977	97.81	882.9	0.971	+1
Strontium	Sr	38	87.62	769	1384	2.54	+2
Sulfur	S	16	32.06	112.8	444.7	2.07	-2, +4, +6
Tantalum	Ta	73	180.9479	2996	5425	16.654	+5
Technetium	Tc	43	(98)	2172	4877	11.5	+4, +6, +7
Tellurium	Te	52	127.6	449.5	989.8	6.24	-2, -4, +6
Terbium	Tb	65	158.9254	1356	3123	8.229	+3
Thallium	Tl	81	204.383	303.5	1457	11.85	+1, +3
Thorium	Th	90	232.0381	1750	4790	11.72	+4
Thulium	Tm	69	168.9342	1545	1947	9.321	+3
Tin	Sn	50	118.69	231.968	2270	7.31	+2, +4
Titanium	Ti	22	47.88	1660	3287	4.54	+2, +3, +4
Tungsten	W	74	183.85	3410	5660	19.3	+6
Uranium	U	92	238.0289	1132.3	3818	18.95	+3, +4, +5, +6
Vanadium	V	23	50.9415	1890	3380	6.11	+2, +3, +4, +5
Xenon	Xe	54	131.29	-111.9	-107.1	0.005887	
Ytterbium	Yb	70	179.04	819	1194	6.965	+2, +3

Yttrium	Y	39	88.9059	1522	3338	4.469	+3
Zinc	Zn	30	65.038	419.58	907	7.133	+2
Zirconium	Zr	40	91.22	1852	4377	6.506	+4
Element 104	(Rf)	104					
Element 105	(Ha)	105					
Element 106		106					
Element 107		107					
Element 109		109					

APPENDIX

A. Density and Boiling Points of Some Common Gases
B. Table of Solubilities in Water
C. Selected Polyatomic Ions
D. Solubility Curves
E. Ionization Energies and Electronegativities
F. Relative Strengths of Acids in Aqueous Solution at 1 atm and 298 K
G. Constants for Various Equilibria at 1 atm and 298 K
H. Standard Electrode Potentials
I. Physical Constants
J. Vapor Pressure of Water
K. Standard Energies of Formation of Compounds at 1 atm and 298 K
L. Heats of Reaction at 1 atm and 298 K
M. Selected Radioisotopes
N. Symbols Used in Nuclear Chemistry
O. A Table of Common Logarithms
P. Periodic Table of the Elements
Q. Boiling and Freezing Point of Water
R. Molal Freezing and Boiling Point Constants
S. Thermodynamic Properties

A.

DENSITY AND BOILING POINTS OF SOME COMMON GASES			
Name		Density grams/liter STP*	Boiling Point (at 1 atm) K
Air	—	1.29	—
Ammonia	NH_2	0.771	240
Carbon dioxide	CO_2	1.98	195
Carbon Monoxide	CO	1.25	82
Chlorine	Cl_2	3.21	172
Hydrogen	H_2	0.0899	20
Hydrogen chloride	HCl	1.64	188
Hydrogen sulfide	H_2S	1.54	212
Methane	CH_4	0.714	109
Nitrogen	N_2	1.25	77
Nitrogen (II) oxide	NO	1.34	121
Oxygen	O_2	1.43	90
Sulfur dioxide	SO_2	2.93	263
*STP is defined as 273 K or 0°C and 1 atm or 760 torr			

B.

TABLE OF SOLUBILITIES IN WATER											
i - nearly insoluble ss - slightly soluble s - soluble d - decomposes n - not isolated	acetate	bromide	carbonate	chloride	chromate	hydroxide	iodide	nitrate	phosphate	sulfate	sulfide
Aluminum	ss	s	n	s	n	i	s	s	i	s	d
Ammonium	s	s	s	s	s	s	s	s	s	s	s
Barium	s	s	i	s	i	s	s	s	i	i	d
Calcium	s	s	i	s	s	ss	s	s	i	ss	d
Copper II	s	s	i	s	i	i	d	s	i	s	i
Iron II	s	s	i	s	n	i	s	s	i	s	i
Iron III	s	s	n	s	i	i	s	s	i	ss	d
Lead	s	ss	i	ss	i	i	ss	s	i	i	i
Magnesium	s	s	i	s	s	i	s	s	i	s	d
Mercury I	ss	i	i	i	ss	n	i	s	i	ss	i
Mercury II	s	ss	i	s	ss	i	i	s	i	d	i
Potassium	s	s	s	s	s	s	s	s	s	s	s
Silver	ss	i	i	i	ss	n	i	s	i	ss	i
Sodium	s	s	s	s	s	s	s	s	s	s	s
Zinc	s	s	i	s	s	i	s	s	i	s	i

C.

SELECTED POLYATOMIC IONS			
CH_3COO^-	acetate	MnO_4^-	permanganate
CN^-	cyanide	MnO_4^{2-}	manganate
CO_3^{2-}	carbonate	NH_4^+	ammonium
HCO_3^-	hydrogen carbonate	NO_2^-	nitrite
$C_2O_4^{2-}$	oxalate	NO_3^-	nitrate
ClO^-	hypochlorite	OH^-	hydroxide
ClO_2^-	chlorite	PO_4^{3-}	phosphate
ClO_3^-	chlorate	SCN^-	thiocyanate
ClO_4^-	perchlorate	SO_3^{2-}	sulfite
CrO_4^{2-}	chromate	SO_4^{2-}	sulfate
$Cr_2O_7^{2-}$	dichromate	HSO_4^-	hydrogen sulfate
Hg_2^{2+}	mercury (I)	$S_2O_3^{2-}$	thiosulfate

D.

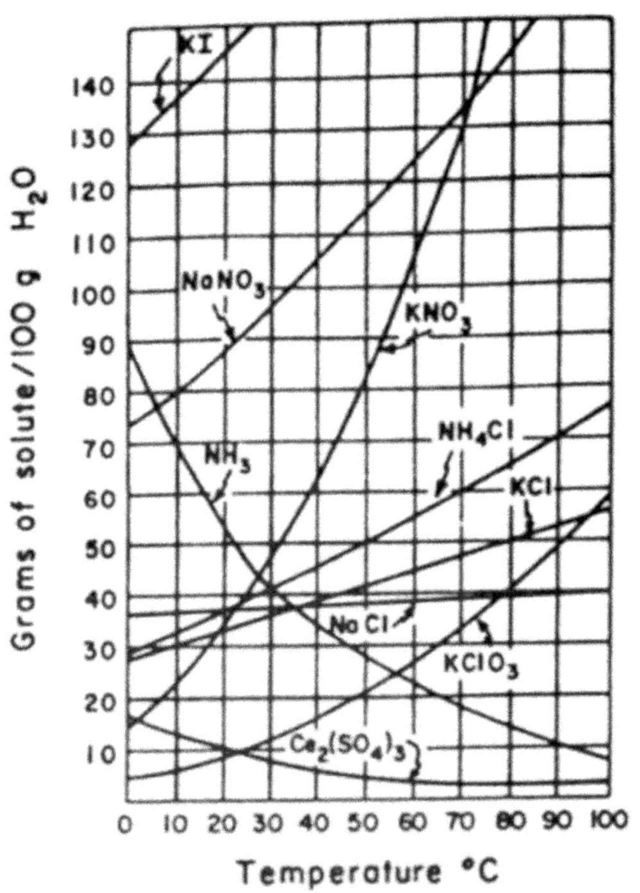

E.

Ionization Energies and Electronegativities							
IA	IIA	IIIA	IVA	VA	VIA	VIIA	0
313 H 2.1	← First Ionization Energy (kcal/mole of atoms) ← Electronegativity						567 He
124 Li 1.0	215 Be 1.5	191 B 2.0	260 C 2.5	336 N 3.0	314 O 3.5	402 F 4.0	497 Ne
119 Na 0.9	176 Mg 1.2	138 Al 1.5	188 Si 1.8	254 P 2.1	239 S 2.5	300 Cl 3.0	363 Ar
100 K 0.8	141 Ca 1.0	138 Ga 1.6	187 Ge 1.8	231 As 2.0	225 Se 2.4	273 Br 2.8	323 Kr
96 Rb 0.8	131 Sr 1.0	133 In 1.7	169 Sn 1.8	199 Sb 1.9	208 Te 2.1	241 I 2.5	280 Xe
90 Cs 0.7	120 Ba 0.9	141 Tl 1.8	171 Pb 1.8	185 Bi 1.9	Po 2.0	At 2.2	248 Rn
Fr 0.7	Ra						

F.

RELATIVE STRENGTHS OF ACIDS IN AQUEOUS SOLUTION AT 1 atm AND 298 K		
Conjugate Pairs		K_a
ACID	BASE	
$HI = H^+ + I^-$		very large
$HBr = H^+ + Br^-$		very large
$HCl = H^+ + Cl^-$		very large
$HNO_3 = H^+ + NO_3^-$		very large
$H_2SO_4 = H^+ + HSO_4^-$		large
$H_2O + SO_2 = H^+ + HSO_3^-$		1.5×10^{-2}
$HSO_4^- = H^+ + SO_4^{2-}$		1.2×10^{-2}
$H_3PO_4 = H^+ + H_2PO_4^-$		7.5×10^{-3}
$Fe(H_2O)_6^{3+} = H^+ + Fe(H_2O)_5(OH)^{2+}$		8.9×10^{-4}
$HNO_2 = H^+ + NO_2^-$		4.6×10^{-4}
$HF = H^+ + F^-$		3.5×10^{-4}
$Cr(H_2O)_6^{3+} = H^+ + Cr(H_2O)_5(OH)^{2+}$		1.0×10^{-4}
$CH_3COOH = H^+ + CH_3COO^-$		1.8×10^{-5}
$Al(H_2O)_6^{3+} = H^+ + Al(H_2O)_5(OH)^{2+}$		1.1×10^{-5}
$H_2O + CO_2 = H^+ + HCO_3^-$		4.3×10^{-7}
$HSO_3^- = H^+ + SO_3^{2-}$		1.1×10^{-7}
$H_2S = H^+ + HS^-$		9.5×10^{-8}
$H_2PO_4^- = H^+ + HPO_4^{2-}$		6.2×10^{-8}
$NH_4^+ = H^+ + NH_3$		5.7×10^{-10}
$HCO_3^- = H^+ + CO_3^{2-}$		5.6×10^{-11}
$HPO_4^{2-} = H^+ + PO_4^{3-}$		2.2×10^{-13}
$HS^- = H^+ + HS^{2-}$		1.3×10^{-14}
$H_2O = H^+ + OH^-$		1.0×10^{-14}
$OH^- = H^+ + O^{2-}$		$< 10^{-36}$
$NH_3 = H^+ + NH^{2-}$		very small

G.

SOME EQUILIBRIUM SYSTEMS AND CONSTANTS		
REACTION	EQUILIBRIUM EXPRESSION	K_{eq}
$2NO_2(g) \rightleftarrows N_2O_4(g)$	$K_{eq} = \frac{[N_2O_4]}{[NO_2]^2}$	1.20 at 55°C
$N_2(g) + 3H_2(g) \rightleftarrows 2NH_3(g)$	$K_{eq} = \frac{[NH_3]^2}{[N_2][H_2]^3}$	625 at 200°C
$2HI \rightleftarrows H_2(g) + I_2(g)$	$K_{eq} = \frac{[H_2][I_2]}{[HI]^2}$	1.85×10^{-2} at 425°C 85 at 25°C
$2SO_2(g) + O_2(g) \rightleftarrows 2SO_3(g)$	$K_{eq} = \frac{[SO_3]^2}{[SO_2]^2[O_2]}$	261 at 727°C
$PCl_5(g) \rightleftarrows PCl_3(g) + Cl_2(g)$	$K_{eq} = \frac{[PCl_3]][Cl_2]}{[PCl_3]}$	2.24 at 227°C 33.3 at 487°C
$COCl_2(g) \rightleftarrows CO(g) + Cl_2(g)$	$K_{eq} = \frac{[CO]][Cl_2]}{[COCl_2]}$	8.2×10^{-2} at 627°C
$2NO(g) + O_2 \rightleftarrows 2NO_2(g)$	$K_{eq} = \frac{[NO_2]^2}{[NO_2]^2[O_2]}$	6.45×10^5 at 227°C
$C(s) + 2H_2(g) \rightleftarrows CH_4(g)$	$K_{eq} = \frac{[CH_4]}{[H_2]^2}$	8.1×10^3 at 25°C
$CO(g) + H_2O(g) \rightleftarrows CO_2(g) + H_2(g)$	$K_{eq} = \frac{[CO_2][H_2]}{[CO][H_2O]}$	1.02×10^3 at 25°C 10.0 at 690°C 3.59 at 800°C
$H_2(g) + Cl_2(g) \rightleftarrows 2HCl(g)$	$K_{eq} = \frac{[HCl]^3}{[H_2][Cl_2]}$	1.8×10^{33} at 25°C
$C(s) + H_2O(g) \rightleftarrows Co(g) + H_2(g)$	$K_{eq} = \frac{[CO][H_2]}{[H_2O]}$	1.96 at 1000°C

SOME SOLUBILITY PRODUCT CONSTANTS AT 25°C			
COMPOUND	K_{eq}	COMPOUND	K_{eq}
AgCl	1.7×10^{-10}	$SrCrO_4$	3.6×10^{-5}
AgBr	5.0×10^{-13}	$BaCrO_4$	8.5×10^{-11}
AgI	8.5×10^{-17}	$PbCrO_4$	2×10^{-16}
$AgBrO_3$	5.4×10^{-3}	$CaSO_4$	2.4×10^{-3}
$AgIO_3$	3.1×10^{-3}	$SrSO_4$	7.6×10^{-7}
		$PbSO_4$	1.3×10^{-8}
		$BaSO_4$	1.6×10^{-9}
		$RaSO_4$	4×10^{-11}

H.

STANDARD ELECTRODE POTENTIALS		
Ionic Concentrations 1M Water at 298°K, 1 atm		
Half-reaction		E°
$F_2(g)+2e^-$	$\rightarrow 2F^-$	+2.87
$MnO_4^- + 8H^+ + 5e^-$	$\rightarrow Mn^{2+} + 4H_2O$	+1.52
$Au^{3+} + 3e-$	$\rightarrow Au(s)$	+1.50
$Cl_2(g) + 2e_-$	$\rightarrow 2Cl^-$	+1.36
$Cr_2O_7^{2-} + 14^+ + 6e^-$	$\rightarrow 2Cr^{3+} + 7H_2O$	+1.33
$MnO_2(s) + 4H^+ + 2e^-$	$\rightarrow Mn^{2+} + 2H_2O$	+1.28
$\frac{1}{2} O_2(g) + 2H^+ + 2e^-$	$\rightarrow H_2O$	+1.23
$Br2(l) + 2e^-$	$\rightarrow 2Br^-$	+1.06
$NO_3^- + 4H^+ + 3e^-$	$\rightarrow NO(g) + 2H_2O$	+0.96
$H_2O_2(g) + 2H^+ (10^-7M) + 2e^-$	$\rightarrow H_2O$	+0.82
$Ag^+ + e^-$	$\rightarrow Ag(s)$	+0.80
$\frac{1}{2} Hg_2^{2+} + e^-$	$\rightarrow Hg(l)$	+0.79
$Hg^{2+} + 2e^-$	$\rightarrow Hg(l)$	+0.78
$NO_3^- + 2H^+ + e^-$	$\rightarrow NO_2(g) + H_2O$	+0.78
$Fe^{3+} + e^-$	$\rightarrow Fe^{2+}$	+0.77
$I_2(s) + 2e^-$	$\rightarrow 2I^-$	+0.53
$Cu^+ + e^-$	$\rightarrow Cu(s)$	+0.52
$Cu^{2+} + 2e^-$	$\rightarrow Cu(s)$	+0.34
$SO_4^{2-} + 4H^+ + 2e^-$	$\rightarrow SO_2(g) + 2H_2O$	+0.17
$Sn^{4+} + 2e^-$	$\rightarrow Sn^{2+}$	+0.15
$2H^+ + 2e^-$	$\rightarrow H_2(g)$	0.00
$Pb^{2+} + 2e^-$	$\rightarrow Pb(s)$	-0.13
$Sn^{2+} + 2e^-$	$\rightarrow Sn(s)$	-0.14
$Ni^{2+} + 2e^-$	$\rightarrow Ni(s)$	-0.25
$Co^{2+} + 2e^-$	$\rightarrow Co(s)$	-0.28
$2H^+ (10^{-7}M) + 2e^-$	$\rightarrow H_2(g)$	-0.41
$Fe^{2+} + 2e^-$	$\rightarrow Fe(s)$	-0.44
$Cr^{3+} + 3e^-$	$\rightarrow Cr(s)$	-0.74
$Zn^{2+} + 2e^-$	$\rightarrow Zn(s)$	-0.76
$2H_2O + 2e^-$	$\rightarrow 2OH^- + H_2(g)$	-0.83
$Mn^{2+} + 2e^-$	$\rightarrow Mn(s)$	-1.18
$Al^{3+} + 3e^-$	$\rightarrow Al(s)$	-1.66

$Mg^{2+} + 2e^-$	$\rightarrow Mg(s)$	-2.37
$Na^+ + e^-$	$\rightarrow Na(s)$	-2.71
$Ca^{2+} + 2e^-$	$\rightarrow Ca(s)$	-2.87
$Sr^{2+} + 2e^-$	$\rightarrow Sr(s)$	-2.89
$Ba^{2+} + 2e^-$	$\rightarrow Ba(s)$	-2.90
$Cs^+ + e^-$	$\rightarrow Cs(s)$	-2.92
$K^+ + e^-$	$\rightarrow K(s)$	-2.92
$Rb^+ + e^-$	$\rightarrow Rb(s)$	-2.93
$Li^+ + e^-$	$\rightarrow Li(s)$	-3.00

I.

PHYSICAL CONSTANTS		
NAME	*SYMBOL*	*VALUES*
Speed of light	c	3.00×10^4 meters/sec
Avogadro number	N_A	6.02×10^{23} per moles
Universal gas Constant	R	0.0821 liter•atm/mole•K 1.99 cal/mol•K 8.32 joule/mole•K
Planck's Constant	h	6.63×10^{-24} joule•sec 1.58×10^{-37} kcal•sec
Charge of electron	e	1.60×10^{-19} coulomb
Molal freezing point depression constant for H_2O = 1.86°C		
Molal boiling point elevation constant for H_2O = 0.52°C		
Atomic Mass unit	1 amu = 1.66×10^{-24} g	
Heat Equivalent	1 kcal = 4.19×10^2 joule	
Volume Standard	1 liter = 1.00×10^2 cm^3	
Angstrom Unit	1 Å = 1.00×10^{-10} meter	
Electron Volt	1 ev = 1.60×10^{-19} joule	

J.

VAPOR PRESSURE OF WATER			
°C	Torr (mm Hg)	°C	Torr (mm Hg)
0	4.6	26	25.2
5	6.5	27	26.7
10	9.2	28	28.3
15	12.8	29	30.0
16	13.6	30	31.8
17	14.5	40	55.3
18	15.5	50	92.5
19	16.5	60	149.4
20	17.5	70	233.7
21	18.7	80	355.1
22	19.8	90	525.8
23	21.1	100	760.0
24	22.4	105	906.1
25	23.8	110	-1074.6

K.

Standard Energies of Formation of Compounds at 1 atm and 298 K		
Compound	Heat (Enthalpy) of Formation Kcal/mole (ΔHf)	Free Energy of Formation Kcal/mole (ΔGf)
Aluminum oxide Al_2O3 (s)	-399.1	-376.8
Ammonia NH3 (g)	-11.0	-4.0
Barium sulfate $BaSO_4$ (s)	-350.2	-323.4
Calcium hydroxide $Ca(OH)_2$ (s)	-235.8	-214.3
Carbon dioxide CO_2 (g)	-94.1	-94.3
Carbon monoxide CO (g)	-26.4	-32.8
Copper (II) sulfate $CuSO_4$ (s)	-184.0	-158.2
Ethane C_2H_6 (g)	-20.2	-7.9
Ethene C_2H_4 (g)	12.5	16.3
Ethyne (acetylene) C_2H_2 (g)	54.2	50.0
Hydrogen fluoride HF (g)	-64.2	-64.7
Hydrogen iodide HI (g)	6.2	0.3
Iodine chloride ICl (g)	4.2	-1.3
Lead (II) oxide MgO (s)	-52.4	-45.3

Magnesium oxide MgO (s)	-143.8	-136.1
Nitrogen (II) oxide NO (g)	21.6	20.7
Nitrogen (IV) oxide NO_2 (g)	8.1	12.4
Potassium chloride KCl (s)	-104.2	-97.6
Sodium chloride NaCl (s)	-98.2	-91.8
Suldure dioxide SO_2 (g)	-71.0	-71.8
Water H_2O (g)	-57.8	-54.6
Water H_2O (l)	-68.3	-56.7
Sample equation 2 Al (s) + O_2 (g) → Al_2O_3 (s)		

L.

Heats of Reaction at 1 atm and 298 K	
Reaction	ΔH (kcal)
CH_4 (g) + $2O_2$ (g) → CO_2 (g) + $2H_2O$ (l)	-212.8
C_3H_4 (g) + $5O_2$ (g) → $3CO_2$ (g) + $4H_2O$ (l)	-530.6
CH_3OH (L) + O_2 (g) → CO_2 (g) + $2H_2O$ (l)	-173.6
$C_6H_{12}O_6$ (s) + $6O_2$ (g) → $6CO_2$ (g) + $6H_2O$ (l)	-669.9
CO (g) + O_2 (g) → CO_2 (g)	-67.7
NaOH (s) Na+ (aq) + OH^- (aq)	-10.6
NH_4Cl (s) NH_4^+ (aq) + Cl^- (aq)	+3.5
H^+ (aq) + OH^- (aq) → H2) (l)	-13.8

M.

SELECTED RADIOISOTOPES		
Nuclide	*Half-life*	*Particle Emission*
^{14}C	5730 y	β^-
^{60}Co	5.3 y	β^-
^{127}Cs	30.23 y	β^-
^{220}Fr	27.5 s	⊠
^{3}H	12.26 y	β^-
^{131}I	8.07 d	β^-

^{40}K	1.28 x 10^9 y	β^+
^{42}K	12.4 h	β^-
^{32}P	14.3 d	β^-
^{224}Ra	1600 y	α
^{90}Sr	28.1 y	β^-
^{238}U	7.1 x 10^4 y	α
^{238}U	4.51 x 10^9 y	α
y = years; d = days; h = hours; s=seconds		

N.

SYMBOLS USED IN NUCLEAR CHEMISTRY		
electron	$^{0}_{-1}e$	β^-
positron	$^{0}_{+1}e$	β^+
proton	$^{1}_{1}H$	p
alpha particle	$^{4}_{2}He$	α
Neutron	$^{1}_{0}n$	n
Gamma radiation		γ

O.

A Table of Common Logarithms

Number	.0	.1	.2	.3	.4	.5	.6	.7	.8	.9
1	.00	.04	.08	.11	.15	.18	.20	.23	.26	.28
2	.30	.32	.34	.36	.40	.40	.42	.43	.45	.46
3	.48	.49	.51	.52	.54	.54	.56	.57	.58	.59
4	.60	.61	.62	.63	.65	.65	.66	.67	.68	.69
5	.70	.71	.72	.72	.74	.74	.75	.76	.76	.77
6	.78	.79	.79	.79	.81	.81	.82	.83	.83	.84
7	.85	.85	.86	.86	.88	.88	.88	.89	.89	.90
8	.90	.91	.91	.92	.92	.93	.93	.94	.94	.95
9	.95	.96	.96	.97	.97	.98	.93	.99	.99	1.00

P.

Periodic Table of the Elements

1 IA 1A	2 IIA 2A	3 IIIB 3B	4 IVB 4B	5 VB 5B	6 VIB 6B	7 VIIB 7B	8	9 VIII 8	10	11 IB 1B	12 IIB 2B	13 IIIA 3A	14 IVA 4A	15 VA 5A	16 VIA 6A	17 VIIA 7A	18 VIIIA 8A
1 **H** Hydrogen 1.008																	2 **He** Helium 4.003
3 **Li** Lithium 6.941	4 **Be** Beryllium 9.012											5 **B** Boron 10.811	6 **C** Carbon 12.011	7 **N** Nitrogen 14.007	8 **O** Oxygen 15.999	9 **F** Fluorine 18.998	10 **Ne** Neon 20.180
11 **Na** Sodium 22.990	12 **Mg** Magnesium 24.305											13 **Al** Aluminum 26.982	14 **Si** Silicon 28.086	15 **P** Phosphorus 30.974	16 **S** Sulfur 32.066	17 **Cl** Chlorine 35.453	18 **Ar** Argon 39.948
19 **K** Potassium 39.098	20 **Ca** Calcium 40.078	21 **Sc** Scandium 44.956	22 **Ti** Titanium 47.867	23 **V** Vanadium 50.942	24 **Cr** Chromium 51.996	25 **Mn** Manganese 54.938	26 **Fe** Iron 55.845	27 **Co** Cobalt 58.933	28 **Ni** Nickel 58.693	29 **Cu** Copper 63.546	30 **Zn** Zinc 65.38	31 **Ga** Gallium 69.723	32 **Ge** Germanium 72.631	33 **As** Arsenic 74.922	34 **Se** Selenium 78.971	35 **Br** Bromine 79.904	36 **Kr** Krypton 83.798
37 **Rb** Rubidium 85.468	38 **Sr** Strontium 87.62	39 **Y** Yttrium 88.906	40 **Zr** Zirconium 91.224	41 **Nb** Niobium 92.906	42 **Mo** Molybdenum 95.95	43 **Tc** Technetium 98.907	44 **Ru** Ruthenium 101.07	45 **Rh** Rhodium 102.906	46 **Pd** Palladium 106.42	47 **Ag** Silver 107.868	48 **Cd** Cadmium 112.414	49 **In** Indium 114.818	50 **Sn** Tin 118.711	51 **Sb** Antimony 121.760	52 **Te** Tellurium 127.6	53 **I** Iodine 126.904	54 **Xe** Xenon 131.294
55 **Cs** Cesium 132.905	56 **Ba** Barium 137.328	57-71	72 **Hf** Hafnium 178.49	73 **Ta** Tantalum 180.948	74 **W** Tungsten 183.84	75 **Re** Rhenium 186.207	76 **Os** Osmium 190.23	77 **Ir** Iridium 192.217	78 **Pt** Platinum 195.085	79 **Au** Gold 196.967	80 **Hg** Mercury 200.592	81 **Tl** Thallium 204.383	82 **Pb** Lead 207.2	83 **Bi** Bismuth 208.980	84 **Po** Polonium [208.982]	85 **At** Astatine 209.987	86 **Rn** Radon 222.018
87 **Fr** Francium 223.020	88 **Ra** Radium 226.025	89-103	104 **Rf** Rutherfordium [261]	105 **Db** Dubnium [262]	106 **Sg** Seaborgium [266]	107 **Bh** Bohrium [264]	108 **Hs** Hassium [269]	109 **Mt** Meitnerium [278]	110 **Ds** Darmstadtium [281]	111 **Rg** Roentgenium [280]	112 **Cn** Copernicium [285]	113 **Nh** Nihonium [286]	114 **Fl** Flerovium [289]	115 **Mc** Moscovium [289]	116 **Lv** Livermorium [293]	117 **Ts** Tennessine [294]	118 **Og** Oganesson [294]

Series															
Lanthanide Series	57 **La** Lanthanum 138.905	58 **Ce** Cerium 140.116	59 **Pr** Praseodymium 140.908	60 **Nd** Neodymium 144.243	61 **Pm** Promethium 144.913	62 **Sm** Samarium 150.36	63 **Eu** Europium 151.964	64 **Gd** Gadolinium 157.25	65 **Tb** Terbium 158.925	66 **Dy** Dysprosium 162.500	67 **Ho** Holmium 164.930	68 **Er** Erbium 167.259	69 **Tm** Thulium 168.934	70 **Yb** Ytterbium 173.055	71 **Lu** Lutetium 174.967
Actinide Series	89 **Ac** Actinium 227.028	90 **Th** Thorium 232.038	91 **Pa** Protactinium 231.036	92 **U** Uranium 238.029	93 **Np** Neptunium 237.048	94 **Pu** Plutonium 244.064	95 **Am** Americium 243.061	96 **Cm** Curium 247.070	97 **Bk** Berkelium 247.070	98 **Cf** Californium 251.080	99 **Es** Einsteinium [254]	100 **Fm** Fermium 257.095	101 **Md** Mendelevium 258.1	102 **No** Nobelium 259.101	103 **Lr** Lawrencium [262]

Q.

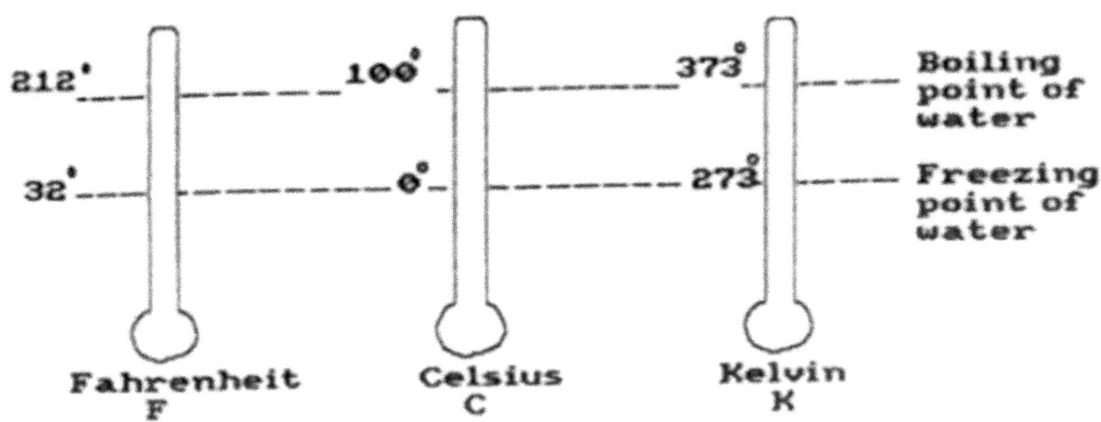

R.

MOLAL FREEZING POINT AND BOILING POINT CONSTANTS				
SOLVENT	FREEZING POINT	K_f	BOILING POINT	k_b
Water	0°C	1.86	100°C	0.52
Acetic acid	17°C	3.90	118°C	2.93
Benzene	5.50°C	5.10	80.0°C	2.53
Cyclohexane	6.5°C	20.2	81°C	2.79
Camphor	178°C	40.0	208°C	5.95
p-Dichlorobenzene	53°C	7.1	—	—

S

Thermodynamics Properties (25°°, 100.000 kPa)			
$\Delta H_t°$ in kJ/mol $\Delta G_t°$ in kJ/mol S° in J/mol•K (concentration of aqueous solutions is 1*M*)			
Substance	**$\Delta H_t°$**	**$\Delta G_t°$**	**S°**
Ag(cr)	0	0	42.55
Ag_2SO_4(aq)	-698	-590	33.1
Al(cr)	0	0	28.33
Al_2O_3(cr)	-1675.7	-1582.3	50.92
Br_2(g)	30.9	3.1	245.4
Br_2(l)	0	0	152.231
CH_4(g)	-74.81	-50.72	186.264
C_2H_4(g)	52.26	68.15	219.56
C_2H_6(g)	-84.7	-32.9	229
C_4H_{10}(g)	-125	-15.7	310
CO_2(g)	-393.509	-394.359	213.74
Ca(cr)	0	0	41.42
$CaCl_2$(aq)	-878	-815	54.8
CaO(cr)	-635	-604	38.2
$Ca(OH)_2$(cr)	-986.09	-898.49	77.41
Cl_2(g)	0	0	223
Cu(cr)	0	0	33.1
$Cu(NO_3)_2$(aq)	-350	-157	193
CuSO4(aq)	-679	-877	109
Fe(cr)	0	0	27.3
Fe_2O_3(cr)	-824.2	-742.2	87.40
H(g)	217.965	203	114.713
H_2 (g)	0	0	130.684
HBr(g)	-36.40	-53.45	198.695
HCl(g)	-92.307	-95.299	186.908
HCl(aq)	-167.159	-131.228	56.5
HNO_2(aq)	-119	-55.6	46.1
HNO_3(aq)	-207	-111	53.3
H_2O(l)	-285.830	-237.129	69.91
H_2O(g)	-241.818	-228.572	188.825
H_2O_2(l)	-186	-120.35	109.6

H_3PO_4(aq)	-1279.0	-1119.1	110.50
H_2SO_4(l)	-814	-690	139
H2SO4(aq)	-909.27	-744.53	20.1
KBr(cr)	-393.798	-380.66	95.90
K_2SO_4(aq)	-1409	—	—
Mg(cr)	0	0	32.5
Mg_3N_2(cr)	-461	-422	88
$Mg(NO_3)_2$(aq)	-875	-677	175
$Mg(OH)_2$(cr)	-925	-834	63.1

GLOSSARY

Absolute zero: The lowest temperature possible (never reached), O K.

Accuracy: A term that describes how close a measurement is to its true or accepted value.

Acetone: Propanone, the first and simplest member of the ketone group.

Acetylene: Ethyne, CH2, the first member of the alkyne series. **Acid:** Any species donate a proton to another species.

Activated complex: An unstable intermediate product of a chemical reaction.

Activation energy: The energy required to form the activated complex and initiate a chemical reaction.

Addition polymerization: The result of joining monomers of unsaturated compounds, usually hydrocarbons, by "opening" double or triple bonds in the carbon chain.

Addition reaction: A reaction that occurs when one or more atoms are added to an unsaturated hydrocarbon molecule at a double or triple bond Alcohol: A hydrocarbon in which one or more hydrogen atoms have been replaced by an—OH group.

Aldehydes: A group of organic compounds represented by the general formula R---CHO, where R represents one hydrogen atom or any hydrocarbon group.

Alkadiene: A hydrocarbon that has two double bonds and the general formula C_nH_{2n-n}.

Alkali metals: The elements of Group 1 of the Periodic Table (the first vertical column).

Alkaline earth metals: The elements of Group 2 of the Periodic table (the second vertical column).

Alkane (methane) series: A group of saturated hydrocarbons with the general formula C_nH_{2n+n}.

Alkene (olefin) series: A group of unsaturated hydrocarbons with one carbon-carbon double bond and the general formula C_nH_{2n}.

Alkyne series: A group of unsaturated hydrocarbons with one triple bond and the general formula C C_nH_{2n-n}.

Alpha decay: The emission of alpha particles, the nuclei of helium atoms, from the nuclei of naturally occurring radioactive isotopes.

Alpha particle: A radioactive emanation that has the same composition as the nucleus of a helium atom, two protons and two neutrons.

Amphoteric (or amphiprotic): A substance that can act either as an acid or as a base. In the Bronsted-Lowry system, amphoteric substances are listed as both acids and bases.

Anion: A negatively charged ion.

Anode: In any electrochemical cell, the electrode at which oxidation occurs.

Aromatic Hydrocarbon: Another name for members of the benzene series.

Artificial transmutation: The bombardment of the nucleus of an atom with high-energy particles such as protons, neutrons, and alpha particles in order to make the nucleus artificially radioactive.

Atom: The smallest unit of all matter.

Atom mass unit (amu): The mass of an atom as given in AMU; the atomic mass of carbon-12 is defined as 12 amu.

Atomic number: The number of protons in the nucleus of an atom.

Average kinetic energy: The temperature of a body.

Avogadro's Hypothesis: Equal volumes of all gases, measured at the same temperature and pressure, contain the same number of particles.

Avogadro's number: The number of atoms in 1 gram atomic mass of any element (approximately 6.02×10^{23}) or the number of molecules in 1 gram molecular mass (1 mole) of any compound.

Base: Any species (molecule or ion) that can combine with (accept) a proton

Benzene series: A group of hydrocarbons with the general formula C_2H_{2n-6}.

Beta decay: The emission of beta particles, or high speed electrons, from the nucleus of naturally occurring radioactive isotopes, increasing the atomic number by 1.

Beta particle: A high-speed electron emitted in many radioactive processes.

Binary compound: A compound containing two elements and whose name usually ends in -ide.

Boiling point: The temperature at which a liquid boils. At the boiling point, the vapor pressure of the liquid is equal the atmospheric pressure.

Boiling point elevation: The elevation of the boiling point of a solvent due to the presence of a nonvolatile solute and proportional to the concentration of dissolved solute particles.

Boyle's law: At constant temperature, the volume of a given mass of gas varies inversely with the pressure exerted on it.

Breeder reactor: A type of nuclear reactor that produces both energy and plutonium-239, a nuclear fuel. The reactor produces more fuel than it consumes.

Bronsted-Lowry acid (or base): A proton donor (or acceptor).

Calorie: The quantity of energy required to raise the temperature of 1 gram of water 1°C

Calorimeter: A device that measures heat absorbed or released in a chemical reaction.

Carbon dating: A procedure in which radioactive isotopes are used to determine the age of ancient objects.

Carboxylic acid: An organic acid with the general formula R-COOH.

Catalyst: A substance that increases the rate of reaction without itself being chemically altered.

Cathode: In any electrochemical cell, the electrode at which reduction occurs.

Cation: A positively charged ion.

Celsius scale: A temperature scale with 100 degrees between the melting point of ice, defined as 0°, and the boiling point of water, defined as 100°.

Chain reaction: A fission reaction in which the products (neutrons) continue to sustain the reaction.

Charles' law: At constant pressure, the volume of a given mass of gas varies directly with the Kelvin (absolute) temperature.

Chemical cell: See Electrochemical cell. **Chemical change:** A change in which new materials or substances are formed.

Chemical equation: A statement employing chemical symbols that describes the qualitative and quantitative relationships in a chemical reaction.

Chemical kinetics: The branch of chemistry concerned with the rates at which chemical retains occur and the physical mechanisms, or pathways, along which they proceed. **Chemistry:** The study of the nature of matter and the changes that matter undergoes. **Colligative properties:** Properties that vary with the concentration of solute particles. **Colloid:** A homogeneous mixture containing particles large enough to reflect a beam of light.

Common ion effect: The decrease in ionization of a weak electrolyte by the addition of a salt with an ion common to the solution of the electrolyte.

Compound: A substance that can be decomposed by chemical action into simpler substances.

Concentrated solution: A solution with a relatively large amount of solute compared to the amount of solvent.

Conceptual definition: A definition based on interpretation of observation and empirical evidence. (See also operational definition)

Condensation: An exothermic change from gas to liquid.

Condensation polymerization: The result of the bonding of monomers by a dehydration reaction.

Conjugate pair: In the Bronsted-Lowry theory, an acid and its corresponding base; they differ only by the presence or absence of a transferable proton.

Contact process: A process for making concentrated sulfuric acid.

Control rods: Rods made of boron or cadmium that can be moved in and out of a nuclear reactor to control the rate of a fission reaction.

Coordinate covalent bond: A bond in which both electrons of a shared pair in a covalent bond are donated by the same atom.

Corrosion: A redox reaction between a metal and substances in its surroundings that destroys the metal.

Covalent atomic radius: One-half of the distance between the nuclei of atoms joined by a covalent bond.

Cracking: A process in which large molecules of hydrocarbons are broken down into smaller molecules.

Crystal: A structure in which the particles of a substance are arranged in a regular geometric pattern.

Dalton's law: The total pressure of a gas mixture is equal to the sum of the individual pressures of the gases comprising the mixture.

Daniell cell: An electrochemical cell using copper and zinc half-cells.

Decomposition: A chemical reaction in which a substance is broken down into simpler substances.

Dehydrating agent: removes water

Density: Mass per unit of volume,

Deuterium: An isotope of hydrogen, H. **Dilute solution:** A solution with very little solute compared to the amount of solvent. **Dipole:** An asymmetrical molecule whose centers of positive and negative charge are located at different parts of the molecule.

Dipole-dipole attraction: The force of attraction between dipoles.

Dissociation: The separation of an ionic or covalent compound into simpler species. **Double bond:** A bond involving two shared pairs of electrons.

Double replacement: A chemical reaction in which two binary (or ternary) compounds react to form new compounds.

Ductile substance: A solid material or substance that can be drawn out to form a thin wire.

E°: Standard electrode potential that describes the spontaneity of a redox reaction.

Effectiveness of collisions: The likelihood that a chemical change will occur following a collision of reacting particles.

Electrochemical cell: An arrangement of half-cells producing a flow of electrons electric current).

Electrode: A metallic conductor in any electrochemical cell.

Electrolysis: The decomposition of an electrolyte by an electric current.

Electrolyte: A substance whose aqueous solution conducts electric current.

Electron: The fundamental unit of negative charge.

Electron cloud: The space around the nucleus of an atom where its electrons are found; an orbital.

Electronegativity: A measure of the ability of a nucleus.

Electroplating: The process of coating a metal in an electrolytic cell.

Element: A substance that cannot be decomposed by ordinary chemical means into simpler substances.

Empirical formula: The simplest whole-number ratio in which atoms combined to form a compound.

Endothermic reaction: A reaction that has a positive H in which energy is absorbed and the potential energy of the products is greater than the potential energy of the reactants. **End point:** The point at which neutralization occurs in an acid-base titration.

Energy: The capacity to do work. Energy level: The energy of an electron associated with its location in the charge cloud and its distance from the nucleus of the atom.

Enthalpy: A measure of the potential energy in chemical bonds.

Enthalpy change: H; commonly known as heat of reaction.

Entropy: A measure of the disorder, randomness, or lack of organization in a system.

Enzynme: An organic catalyst.

Equilibrium: A state of balance between two opposing reactions (Physical or chemical) occurring at the same rate.

Equilibrium constant (K_{eg+}): The numerical value of the ratio between the concentrations of the products and the concentrations of the reactants in a system at equilibrium.

Ester: An organic compound formed when an alcohol reacts with a carboxylic acid; represented by the general formula R-COO-R'.

Esterification: The reaction of an organic acid with an alcohol to give an ester and water.

Ether: A hydrocarbon derivative represented by the formula R-O-R'.

Evaporation: A spontaneous change from liquid to gas.

Excited state: The condition of an atom that exists from absorption of energy, characterized by the movement of one or more electrons to higher energy levels.

Exothermic reaction: A reaction that has a negative H in which energy is released and the potential energy of the products is less than the potential energy of the reactants.

Family: See group.

Fermentation: A process in which enzymes act as catalysts to produce alcohol from carbohydrates.

Fission reaction: See nuclear fission.

Fission reactor: A type of nuclear reactor presently in use that produces energy from the fission of uranium.

Fractional distillation: A technique used to separate components of a mixture, based on the differences in their boiling points.

Free energy change: The difference between energy and entropy change.

Freezing point: The temperature at which a liquid and its solid form can coexist in equilibrium.

Freezing point depression: The depression of the freezing point of the solvent due to the presence of a solute and proportional to the concentration of dissolved solute particles. **Fuel (nuclear):** Fissionable isotopes used in nuclear reactors.

Function group: A particular arrangement of atoms associated with certain chemical and physical properties that are characteristic of a series of organic compounds.

Fused: Term for a molten substance that is a solid at room temperature.

Fusion reaction: See nuclear fusion.

Galvanic cell: See electrochemical cell.

Gamma radiation: Short wave length radiation similar to high energy X-rays.

Gas: A phase of matter that takes the shape and volume of its container.

Glycerol: A trihydroxy alcohol.

Glycol: A dihydroxy alcohol containing two- OH groups, such as ethylene glycol.

Graham's law: Under the same conditions of temperature and pressure, gases diffuse at a rate inversely proportional to the square root of their molecular masses.

Gram atomic mass: The atomic mass of an element expressed in grams; also known as one mole of an element.

Gram formula mass: The sum of the atomic masses of all atoms in a molecular formula, expressed in grams.

Ground state: The normal state of an atom with all of its electrons in their lowest available energy levels.

Group: A family of elements represented by a column in the Periodic Table.

Haber process: A commercial process for the production of ammonia from nitrogen and hydrogen.

Half-cell: A strip of metal in contact with its ions.

Half-life: The time required for one-half of the nuclei of a given sample of a radioactive element to undergo decay.

Half-reaction: One-half of a redox reaction, representing either a loss of electrons or a gain of electrons.

Halogen: An element that is a member of group 17 of the Periodic Table.

Halogen derivative: An organic compound that is the product of halogen atoms of a saturated hydrocarbon are replaced by a corresponding number of halogen atoms.

Heat: The form of energy most often associated with chemical change.

Heat of fusion: The amount of energy required to convert one gram of a substance from solid to liquid at its melting point.

Heat of reaction: The difference in heat content between the products and reactants of a reaction; also known as enthalpy change.

Heat of vaporization: The amount of energy required to convert one gram of substance from liquid to vapor at its boiling point.

Heterogeneous substance: A sample of a substance that is not uniform throughout in its properties composition, or phase.

Homogeneous substance: A sample of a substance that is uniform throughout in its properties, composition, or phase.

Homologous series: A group of organic compounds with the same general formula and similar structures and properties.

Hydrated salt: A solid compound containing a definite number of water molecules bonded directly to the solid crystal.

Hydration: The addition of water to a compound or an ion.

Hydrocarbon: A compound that contains atoms of hydrogen and carbon only.

Hydrogen bonding: The intermolecular attraction between the hydrogen on one molecule and an element of small atomic radius and high electronegativity on another molecule.

Hydrogenation: The addition of hydrogen to an unsaturated hydrocarbon or hydrocarbon derivative.

Hydrolysis: The reaction of a salt with water to form a slightly acidic or basic solution.

Ideal gas: An imaginary gas that obeys the gas laws perfectly.

Immiscible: Term for tow liquids that will not mix to form a solution

Increment: The number and kind of atoms by which each member of a homologous series differs from the preceding member. **Indicators:** Organic substances that change color at definite pH values.

Inert (rare or noble) gases: Group 18 elements in the Periodic Table.

Insoluble: Term for a solid or gaseous substance that does not dissolve in a liquid.

Ionic bond: A bond between oppositely charged particles formed when electrons are transferred between atoms.

Ionic radius: An arbitrary designation that describes the size of an ion, or the size of the electron cloud associated with an ion.

Ionic solids: Crystalline solids containing ions. Ionization: The process of forming ions, often by the reaction between solvent and solute molecules.

Ionization constant (K): The numerical value of the equilibrium ration of dissociated ions to undissociated compound.

Ionization energy: The amount of energy required to remove the most loosely bound electron from an atom in the gas phase. **Isomers:** Compounds that have the same molecular formula but different structural formulas.

Isotopes: Atoms with the same atomic number but different atomic masses.

K_a: The ionization constant of a weak acid. **K_{ap}:** The solubility product constant.

K_w: The ion product of water.

Kelvin scale: A temperature scale with 100 degrees between the melting point of ice and the boiling point of water, with 0 degrees representing absolute zero, or the point of zero kinetic energy.

Kernel: Term used to refer to all parts of the atom except the valence electrons.

Ketones: A group of organic compounds having the general formula R--CO-R'.

Kilocalorie: One thousand calories.

Kinetic energy: The energy of a body in motion.

Kinetic theory of gases: A theory that explains the behavior of gases by assuming that they consist of individual molecules in random motion.

Law: A statement of mathematical formula that summaries certain experimental observations.

Lead storage battery: A cell in which the change in oxidation states of lead is used to produce electricity.

Le Chatelier's principle: When a system at equilibrium is subjected to a stress, the system will shift so as to relieve the stress and move to a new equilibrium.

Liquid: A phase of matter that takes the shape but not necessarily the volume of its container.

Malleable: Term for the ability of a metal to be hammered or rolled into thin sheets.

Mass number: The numerical sum of the protons and neutrons in an atomic nucleus. **Matter:** Anything that occupies space and has mass.

Melting point: The temperature at which a solid and its liquid form can coexist in equilibrium.

Meniscus: The curved shape that the surface of a liquid takes in a glass container.

Metal: An element that contains atoms that lose electrons to form positive ions in chemical reactions.

Metalloid: an element with some of the properties of metals and some of the properties of nonmetals.

Miscible: Term for two liquids that can mix to form a solution.

Moderator: A substance that can slow down neutrons, used in nuclear reactors to control fission reactions.

Molal boiling point constant: A constant that can be multiplied by the molality of the solution to determine the boiling point elevation.

Molal freezing point constant: A constant that can be multiplied by the molality of the solution to determine the freezing point depression.

Molality: The concentration of a solution expressed in moles of solute per kilogram of solvent.

Molar volume: The volume of one mole of a gas, usually expressed as 22.4 L at STP. **Molarity:** The concentration of a solution, expressed in moles of solute per liter of solution.

Mole of an element: A fixed quantity containing Avogadro's number (6.02×10^{23}) of atoms and equal to the gram atomic mass of the element.

Mole of a compound: A fixed quantity containing Avogadro's number of molecules and equal to the gram formula mass of the compound.

Molecular formula: A statement of the exact number of each kind of atom in a molecule or compound.

Molecular solids: Solids formed by molecules of covalently bonded atoms.

Molecule: The smallest subdivision of a substance that has all the properties of the substance.

Monohydroxy alcohol's: Hydrocarbon derivatives containing one-OH group per molecule.

Monomer: A small molecule that joins with other identical molecules to form a polymer. (large molecule).

Network solid: Covalently bonded atoms linked in a network that extends throughout a large sample.

Neutralization: A reaction in which one mole of H from an acid combines with one mole of OH from a base to form water.

Neutron: A nuclear particle that does not have an electrical charge.

Nonelectrolyte: A substance whose aqueous solution does not conduct electric current.

Nonmetal: Term for a bond or a molecule with uniform distribution of charge.

Nonpolar bond: A covalent bond in which the electron pair that forms the bond is shared equally by two atoms.

Nonvolatile: Term for a liquid with negligible vapor pressure.

Normal boiling point: The temperature at which the vapor pressure of a liquid equals 1 atmosphere.

Normal freezing point: The temperature at which the liquid will change to a solid at 1 atmosphere.

Nuclear emanation: The release of alpha, beta, or gamma particles from the nucleus of a naturally occurring radioactive isotope. **Nuclear fusion:** The process of combining two light nuclei to form a heavier nucleus and releasing an even greater amount of energy than in a fission reaction.

Nuclear reactor: A device designed to produce controlled nuclear reactions so that energy is released at a steady and predictable rate.

Nucleons: The particles that compose the nucleus of an atom, generally protons and neutrons.

Operational definition: A definition based on observed characteristics, without interpretation of empirical evidence.

Orbital: The average region of most probable electron location.

Organic acid: A hydrocarbon derivative that has the general formula R--COOH.

Organic chemistry: The branch of chemistry dealing with the compounds of carbon.

Oxidation number: The charge that an atom has, or appears to have, when certain rules for assigning charge are used.

Oxidation-reduction: A chemical reaction in which electrons are removed from some species and transferred to others.

Oxidizing agent: An electron acceptor, by accepting one or more electrons, an oxidizing agent causes an other species to become oxidized and is itself reduced.

Particle accelerator: A device using magnetic and electric fields to accelerate charged particles in order to penetrate a target nucleus.

Percent error: A method of describing the accuracy of a numerical result in an experiment.

Periodic law: The properties of the elements are periodic functions of their atomic numbers.

Phase: Term for any one of the three possible physical states of matter; solid, liquid, or gas.

Phase equilibrium: The equilibrium established between two different phases of a substance, such as ice and water, when a phase change is reversible.

Physical change: A change of state or physical property in which no new materials or substances are formed.

Polarity: Term used to describe a molecule in which one end has a slightly positive charge and the other end has a slightly negative charge because of asymmetrical arrangement of polar bonds.

Polar covalent bond: A covalent bond in which the electron pair is not shared equally by the two atoms; generally, a covalent bond in which the electron pair is shared by atoms of different elements.

Polymer: A large molecule composed of many repeating smaller units called monomers.

Polymerization: A process whereby large molecules are formed from smaller molecules. The manufacture of synthetic rubber is an example.

Primary alcohol: A monohydroxy alcohol in which one --OH group is attached to an end carbon of a hydrocarbon chain.

Proton: The fundamental unit of positive electrical charge.

Quantum: A discrete amount of energy absorbed or radiated.

Quantum number: A number used to described the energy of an electron.

Radioactivity: The spontaneous breakdown of an atomic nucleus, yielding particles and radiant energy.

Radioisotope: An isotope with an unstable nucleus that undergoes radioactive decay.

Reaction rate: The rate at which reactants are consumed or products formed in a chemical reaction.

Reducing agent: An electron donor.

Reduction potential: An indicator of how easily a substance can be reduced, compared to a standard. More easily reduced substances have positive reduction potentials, while less easily reduced substances have negative reduction potentials.

Roasting: Heating an ore in air, a process used to convert sulfate and carbonate ores into oxides.

Rule of eight: The atoms in a molecule and their electrons are arranged so that each atom achieves the electron configuration of an inert gas atom, that is, eight electrons.

Salt: An ionic compound that dissociates of form actions other than H--$^{+}$ and anions other than OH^{-}.

Salt bridge: A u-shaped tube containing a solution of an electrolyte that permits the passage of ions between solutions in the two half-cells of an electrochemical cell while preventing the solutions from mixing.

Saponification: The hydrolysis of a fat to produce a soap and glycerin.

Saturated hydrocarbon: A hydrocarbon that contains single carbon-carbon bonds.

Secondary alcohol: An alcohol in which the --OH group is bonded to a carbon atom that, in turn, is bonded to two other carbon atoms. **Shell:** See energy level.

Shielding effect: Weakening of the force between the nucleus and the outermost valence electrons because of the presence of electrons at inner energy levels that act as barriers.

Significant figures: In a measurement, the digits that are certain plus one uncertain digit.

Single covalent bond: a bond formed by the sharing of one pair of electrons by two atoms.

Single replacement: A chemical reaction in which an element reacts with a binary (or ternary) compound to form a new element and a new binary (or ternary) compound.

Soluble: Term for a solid or gaseous substance that can be dissolved in a liquid.

Solute: The substance or substances that are present in lesser proportions in a solution and are dissolved.

Solution: A homogeneous mixture of two or more substances.

Solvent: The substance that is present in greater proportion in a solution and does the dissolving.

Spectator ion: An ion that does not undergo any change in a chemical reaction.

Spectral lines: Characteristic lines produced by radiant energy of a specific frequency emitted when electrons in an atom in the excited state return to lower energy levels.

Spontaneous reaction: A reaction that can occur in nature under a given set of conditions.

Stable: Term for a compound or system at a relatively low energy level and thus less likely to undergo chemical change.

Standard conditions: The standard conditions of temperature and pressure of the measurement of gasses.

Standard electrode potential: The electrical potential of an electrochemical cell measured relative to a standard hydrogen electrode defined as zero.

Standard solution: A solution of known concentration used to determine the unknown concentration of another solution.

Stock system: A system used to name compounds of metals that have more than one possible ionic charge.

Stoichiometry: The study of the quantitative relationships in chemical formulas and equations.

Sublevels: Subdivisions of the principal energy levels in an atom.

Sublimation: A change from the solid phase directly to the gas phase without passing through any apparent liquid phase. Solids that sublime have high vapor pressures and low intermolecular attractions.

Substance: Any variety of matter for which all samples have identical composition and properties.

Supersaturated solution: A solution that contains more solute that its ordinary capacity at a given temperature.

Synthesis: A chemical reaction in which atoms of elements combine to form compounds.

Temperature: A measure of the average kinetic energy of a body.

Ternary compound: A compound consisting of three elements.

Tertiary alcohol: An alcohol in which the --OH group is bonded to a carbon atom that, in turn, is bonded to three additional carbon atoms.

Tetrahedron: A central atom joined to four other atoms, all the covalent bonds having equal angles of 109.5°C between each pair of bonds.

Theory: A statement or picture that tries to account for certain experimental observations.

Thermometer: A device sued to measure temperature, with calibrations based on the fixed freezing point and boiling point of water.

Titration: A technique used to find the concentration of an acid or base of any other kind of solution through mixing of measured volumes.

Tracer: A radioisotope used to follow the course of a chemical reaction without altering the reaction.

Transition elements: Elements found in Groups 3 through 12 of the Periodic Table in which electrons from the tow outermost sublevels may be involved in a chemical reaction. These elements generally exhibit multiple positive ionic charges and oxidation states.

Transmutation: The change of one element into another due to changes in the nucleus caused by radioactive decay or bombardment by subatomic particles.

Trihydroxy alcohol's: Hydrocarbon compounds containing three --OH groups, such as glycerin. Triple covalent bond: A bond formed by three shared pairs of electrons between two atoms.

Tritium: An isotope of hydrogen with formula ^3H.

Unsaturated hydrocarbon: A hydrocarbon that contains at least one carbon-carbon double or triple bond.

Unsaturated solution: A solution that contains less solute that its ordinary capacity at a given temperature.

Valence electrons: Those electrons found in the outermost energy level; most chemical properties of an atom are related to the valence electrons.

Van der Waals forces: Weak forces of attraction that exist between all molecules, especially nonpolar molecules.

Vapor: Term frequently used to refer to the gas phase of a substance that is ordinarily a liquid or solid at room temperature.

Vapor pressure: The pressure exerted by a saturated vapor in a closed system; vapor pressure increases as the temperature of the liquid increases.

Volatile: Term used to describe a liquid or solid that has a high vapor pressure and thus is easily converted to the gas phase.

Voltaic cell: See electrochemical cell.

Weak acid: An acid, such as acetic acid, that is only very slightly ionized in water solution;

Weak base: A base, such as ammonium hydroxide, that is only very slightly ionized in water solution.

SELECTED ESSAYS

Water is an amazing compound. Some of water's properties are as follows:

Hydrogen Bonding

Hydrogen bonding occurs when hydrogen atoms are bonded to small electronegative atoms resulting in an uneven distribution of charges. This results in hydrogen having a positive charge on one side and a negative charge on the other side.

Moreover hydrogen can be compared to a bare proton attracted to the negative end of the electronegative element.

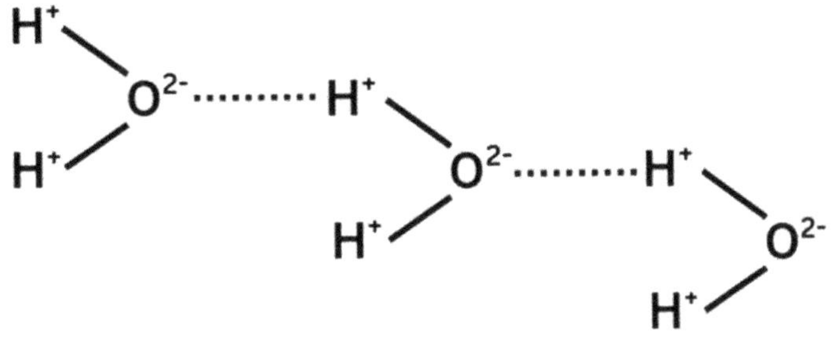

The dashed lines are hydrogen bonds. Note the dipole-dipole charges.

Furthermore ***hydrogen bonding*** suggests that water is a complex molecule because of water's high boiling point. Scientists also believe excess energy is needed to break the complex hydrogen bonds.

Hydrogen bonding accounts for water's high boiling point. When compared to similar compounds with lower boiling points such as H_2S, H_2Te, and H_2S.

Other Compounds with hydrogen bonding occur when hydrogen is bonded to fluorine and nitrogen atoms.

Water

Water is amphoteric substance (has both acid and base properties).

Explanation; The autoionization of water.

$H_2O + H_2O \rightarrow H_2O^+ + OH^-$

acid base

example of acids: *orange juice, vinegar, lemonade*

examples of bases: *shampoo, ammonia, baking soda*

In conclusion water is complex compound with amazing solubility properties, an electrolytic properties and supports living things.

London Dispersion Forces or Van der Waals Forces are weak forces of attraction that exist between nonpolar molecules in solids, liquids and gases especially under high pressure. For when nonpolar molecules in gases approach each other their electrons produce temporarily opposite charged particles.

Furthermore these temporary charged particles influence the boiling points and freezing points of solids. Van der Waals forces increase with an increase in electrons which results in a larger molecular size. Organic chemists believe the boiling points in alkanes and hydrocarbons set a trend because of Van der Waals forces.

Note the temporary charges created between adjacent. H2 molecules.

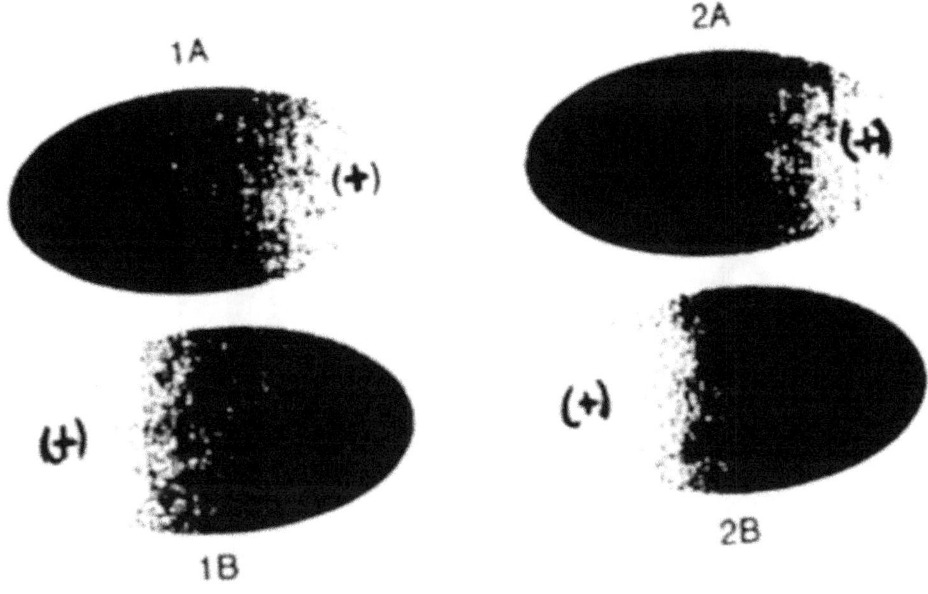

IONIC SOLIDS

Some ionic solids such as sodium chloride, NaCl will not conduct electricity in their solid or crystal state. However when dissolved in water NaCl will conduct a current. This occurs because the attraction between polar water molecules and the solid ions form an ionic solution.

Furthermore most ionic compounds are solids at room temperature. Ionic solids also have high melting points.

Notice the NaCl diagram Na is small spheres and C1 large spheres.

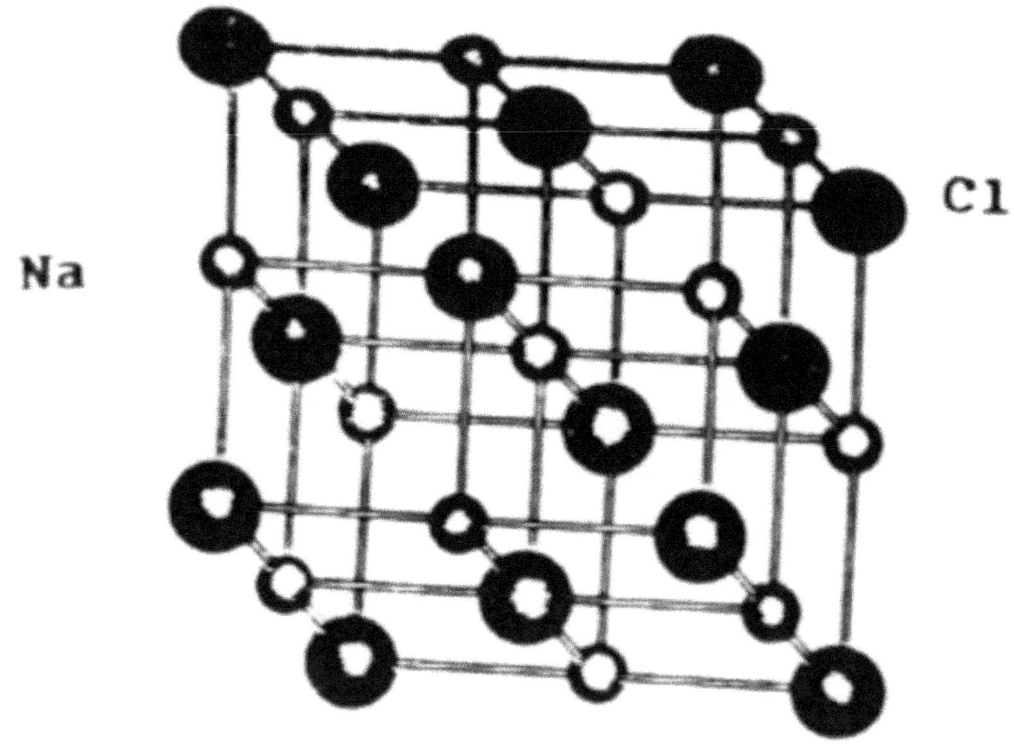

A Topic of Interest

Computers are part of our lives. Our fascinating element used to make computer discs is silicon. Pure silicon is poor conductor of electricity. However when doped or impure an excellent conductor.

Chemists used elements similar to Si, But differ in their valence structure. Arsenic is used in Si and Ge chips because it gives its fifth valence electron. Chemist called a n-type atoms because of its negative charge.

On the other hand if an electron with a slight deficiency is used such as boron, we create a slight deficiency or ***positive hole in the lattice***. The results in electrons move to fill the hole. This creates an excellent conductor.

Note n type Atoms furnish mobile electrons.
p type There is a deficiency of electrons.

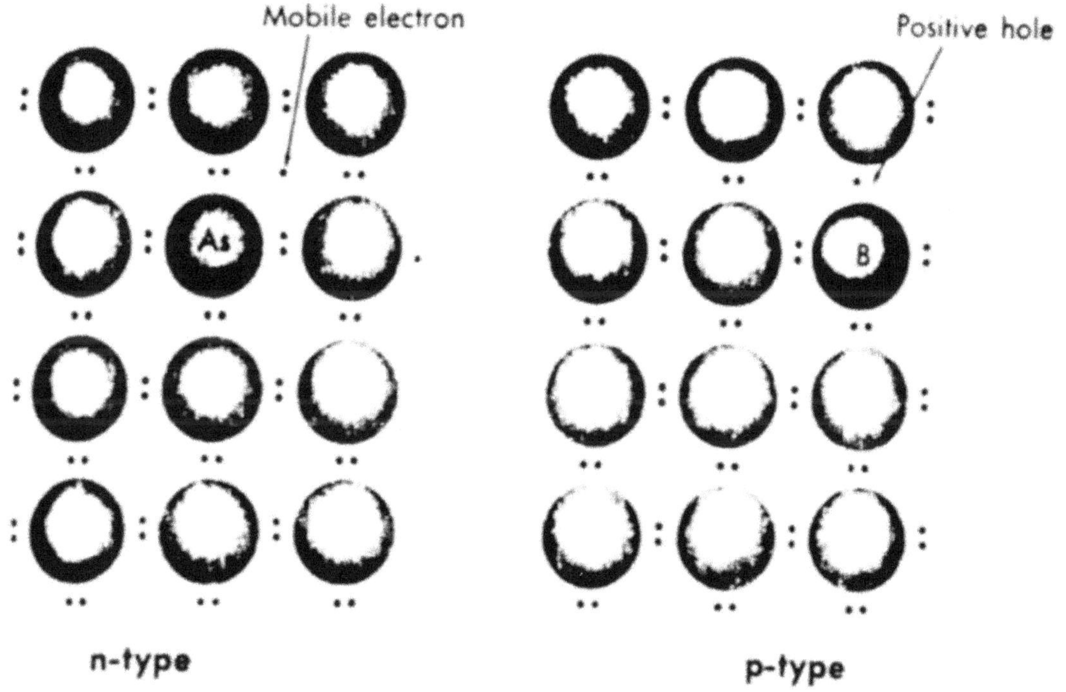

SELECTED REFERENCES

A Few Reference books and Sources

Kolb, D "The Mole" J Chem. Educ.,55, 758 (1978)

Kolb, D "The Chemical Equation, Part 1: Simple Reactions,"

J Chem Educ., 55184 (1978)

Strong, L.E., "Balancing Chemical Equations", Chemistry47 (1), 13 (1974)

Harris. AD "The Density and Apparent Molecular Weight of Air", J Chem Educ., 60 (1984)

The Chemistry of Vat Dyes by Dianne N. Epp 1995

NSTA Publishers #PB056x1

Polymers All Around You by Linda Woodward 1996 NSTA Publishers #op339X

Exploring Chemical Elements and Their Compounds.

David L. Heiserman 1992 NSTA Publishers $OP288X

Chemistry for Every Kid by Janice VanCleave

NSTA Publishers 1989 #OP240X3

The Elements by John Emsley 1991 NSTA Publishers #OP376X

Reactions and Reason by Atkinson and Heikkinen 1991

NSTA Publishers #OP116X

Reviewing Chemistry by Peter Demmin 1996 Amsco Publications

Unique Chemical Labs By Louis Casimier 1994 Stipes Publishing Company champaign Illinois

Solving Problems In Chemistry by Hines Smoot and Smith Merrill publishing Co. 1995

Chemical principles by Masterton and Slowinski Saunders Sunburst College Publishers.

Review Journals of Chemical Education American Chemical Society

Jakuba, Stan "Go Metric Now"

Chem Tech 18 #7 (July 1988) p424-425

Fisher, A Hunting Neutrinos Popular Science 232 #5 (May 1988) pp72-74

Salem, Lionel Marvels of the Molecule New York VCH Publishers 1987

Block, B Peter Inorganic Chemical nomenclature, Wash D.C. ACS, 1990

Chemistry on The Internet (refer to computer address and references)

Garcia, Arcesio "A Method to Balance Chemical Equations" Journal of Chemical Education 63 # 3

March 1987 Pages 247-248

Strong, Judith A. "The Periodic Table and Electron Configuration Journal of Chemical Education 63# 10 (October 1986) pages

ANSWER KEY

Exercise 1.1

1.	104 F	313 A			
2.	2685 C	2958 A			
3.	.038 grams	3.8 x 10^{-5} Kg			
4.	-117.7 C	155.3 A			
5.	.85g	8.5x10^{-4} Kg			
6.	1250 mm	.00125Km or 1.25x10^{-3}			
7.	.264 glml				
8.	49 grams				
9.	28.9 ml				
10.	10,800 grams				
11.	3.25 x 10^{-5}				
12.	.17mg/cc				
13.	42 C	108 F			
14.	7.264 Kg				
15.	231.6				
16.	32,166 ft	6.6 miles	386,000 inches		
17.	141.8 grams				
18.	120 lbs				
19.	.15 grams				
20.	A. 12.5%	B. .75	C. 2.0x10^{-10}	D. 1.7 3250ml	E. 4.35x10
	F. .389	G. .61	H. 3250ml		
21.	2.18x10^{5}				
22.	758cc				

Exercise 2.1

$MgCO_5$ = 84 grams
$C_6H_{12}O_6$ = 180 grams
$KC1O_3$ = 122.5 grams
H_2O= 18
$AgND_3$ = 170
KC1 74.5
$C_{12}H_{22}O_{11}$ = 342
Ba $(NO_3)_2$ = 261.3
Ni $(NO_3)_2$ = 182.7
Bi $(OH)_3$ = 260
Ca $(ClO_3)_2$ = 207
H_2SO_4 = 93
$Na_2Cr_2O_7$ = 262

Fe_2O_3= 160
KCN = 65
H_2CO_3 = 62
$MgSO_4$ = 120
$AsCl_5$ =252.5
$PbSO_4$ = 303
AgCl = 143.5
KOH= 56
$C_6H_8O_6$ =176
N_2H_4CO = 60
CH_3OCH_3 = 46

Chapter 3

A. .098 moles
B. .126 moles
C. 2.11 moles
D. .011 moles
E. .012 holes
F. a. 5.9×10^{22} b. 7.6×10^{22} c. 1.27×10^{24}
G. 3.48 gram
H. 617.025 grams
I. 5360 grams
J. 1.66×10^{-24}
K. 4.45×10^{23}
L. .15 moles
H. 5.37 x 10 molecules
N. .25 moles 8.47 grams
O. .1029 Moles 6.2×10^{22} molecules

Chapter 4 Exercise 4-1

1. 77.4%
2. 48.9%
3. Mg = 41.38%
   O = 55,17%
   H = 3.45%
4. Ca = 40%
   $(NO_3)_2$ = 37.46% Fe_2 = 77.77%
5. .423 moles or (94.1% oxygen)
6. 10.07 grams of Cu
7. 16% Al
8. 9.47 grams of Al
9. 80% C 9.5% H 10.2% O
10. 50.24 grams of C
11. $Na_2S_20_3$
12. 44.1 grams of Au
13. 65% Pt
14. 24% C 76% F
15. C = 32 ml H = 5 ml
16. C = 15.9% H = 2.2% N = 18.5% O = 63.4%
17. C = 53.94% H = 6.4% N = 8.98% S = 10.26% O = 20.5%
18. 21.54 grams of C
19. 3.84 grams of Oxygen
20. C = 38.4% Cl = 56.8 H = 4.8%
21. 8.06 grams of C
    2.238 grams of H
    7.164 grams of O
24. 12.54 grams of N

Chapter 5 Exercise 5-1

A. $Ca\ Br_2$
B. Mn_3S_4
C. $Al_2C_3O_9$
D. NaCl
E. H_2O
F. $CHC1_3$
G. CaO
H. MgN_2
I. Pt_2Cl_5
J. NH_3

Chapter 6 Exercise 6-1

A. $3\ Li_2HPO_4$
B. $2H_3PO_3$
C. $4C_4H_9$
D. $3SiO_2$
E. $4N_2O_5$
F. XY

Chapter 7 questions

Direct Combination 7-1

a $2 + 1 \rightarrow 2$
b $2 + 3 \rightarrow 2$
c $2 + 1 \rightarrow 2$
d $3 + 1 \rightarrow 1$
e $4 + 5 \rightarrow 2$

Displacement

a $1 + 2 \rightarrow 2 + 1$
b $1 + 2 \rightarrow 2 +1$
c $1 + 2 \rightarrow 1 + 1$
d $1 + 2 \rightarrow 1 + 2$
e $1 + 1 \rightarrow 1 + 1$

Decomposition

a $2 \rightarrow 2 + 1$
b $3 \rightarrow 1 + 2 + 1$
c $1 \rightarrow 1 + 2$
d $2 \rightarrow 2 + 3$
e $2 \rightarrow 2 + 3$

Chapter 8

Exercise 8.1

1. $CuSO_4$ $.6\ H_2O$
2. $CoCl_2$ $.2\ H_2O$
3. $MgCo_3$ $.3\ H_2O$
4. Na_2CO_3 $.10\ H_2O$

Chapter 9

1. Because metal activity decrease as we move horizontally.
2. Transitional fill outer shell before inner shell.

Chapter 10

1. 497.35 calories
2. 4800 calories
3. 1094.4 calories
4. 1804 calories
5. 231 calories
6. 0.9498 calories or 3.97 joules

Chapter 11 11-3

A. 2.19m

B. 18.55m

C. .96m

D. 5.02m ZERO 05m

11-4

A. 7.7m

B. 4.41m

C. 5.185m

D. 3.47m

Chapter 12

4 97ml

5. .228m

6. 800ml

7. 2.15m

8. 533.3ml

9. 1.3 ml

10. 3.4 moles

11. 56.6 ml of HNO_3 + 400ml of H_2O

12. 2.4m NaOH

Chapter 13

13-1

1. $1 + 1 + 1 \rightarrow 1 + 2 + 1$
2. $1 + 2 \rightarrow 2 + 1$
3. $4 + 1 \rightarrow 4 + 1$
4. $3 + 4 \rightarrow 3 + 1 + 2$

14. $4 + 5 \rightarrow 4 + 6$

15. $1 + 1 \rightarrow 1 + 2$

13-2

6. $1 + 2 \rightarrow 1 + 1$
7. $1 + 2 \rightarrow 2 + 1$
8. $2 + 2 \rightarrow 2 + 1 + 2$
9. No Equation
10. $6 + 6 \rightarrow 10 + 2 + 6$
11. $1 + 14 \rightarrow 2 + 2 + 7 + 4$
12. $1 + 4 \rightarrow 1 + 2 + 1$
13. $1 + 4 \rightarrow 1 + 1 + 2$

Chapter 14

1. Refer to Test
2. Refer to Test
3. 100.7 grams
4. 326 grams Cu^{+2}
5. 15.4 hours
6. +1.08 volts
7. +3.17 volts
8. +.31 volts
9. +2.37 volts
10. 21.93 amps
11. A = oxidation B = Reduction
12. A = Anode B = Cathode
13. Anode

Chapter 15

1. Refer to Text
2. 215 ml
3. 691 ml
4. 252.4 ml
5. 696 ml
6. 1.84:1
7. 1.4:1
8. 8.5×10^{22}
9. 44 grams

Chapter 15

10. O_2 = 01.423 g/1 N_2 = 1.25 g/l F_2 = 1.7 g/l
11. 875.2 ml
12. 393 ml
13. 1.5 g/1
14. 1.6:1
15. 33.4 grans

Exercise 15-2

A. decreasing
B. .25 moles 1.5×10^{23}
C. Increase
D. Boyles, Charles Lav
E. SO_2 slowest H_2 Fastest
F. 1.5 moles, 25.5 grams
G. 8.03 moles
H. .20 moles 1.22 x .023

Easy Chemistry

16-1

1. .75 moles
2. 2 grams
3. 1.52 moles
4. 1.646 liters
5. 63 grams
6. 156.8 liters
7. 26.88 liters
8. 73 grams
9. 24.5 grams
10. Limited NaCl (0.51 moles)
    Excess H2SO, (1.0 mole)
    60g/117g = 0.51 moles (note 117g=2 moles of NaCl)
    98g/98g (1mole) sulfuric acid

17-1

1. 4.9×10^{-3}
2.
3. 7.7×10^{-13}
   1.0×10^{-10}
4. 1.34×10^{-7}
   moles per liter for Pb. Cr D_4

   4.33×10^{-5}g/l
   AgCl = 1.3×10^{-5} moles per liter
   1.86×10^{-3} g/l

   PbI_2 1.5×10^{-3} moles per liter
   7.0×10^{-1} grams per liter

   $AgC_2H_3O_2$ 3.16×10^{-2} moles per liter

5.27 g./1

17-1

5. Refer to text
6. pressure, temperature, catalyst
7. $K_{sp} = 5.08 \times 10^{-13}$
8. separation, analysis, solubility tests
9. 5.0×10^{-12} is least soluble
10. Temperature

    increase will increase solubility
    decrease will decrease solubility

    Stirring

    will increase solubility

    Similarities

    likes dissolve likes

    Exponents

    + exponents are usually, more soluble
    - exponents are usually less soluble

Exercise 18-2

1. pH = 1
2. pH = 10
3. pH = 1.57
4. pH = 12
5. pH = 1.31

(18-3)

A. pH = 4.05
B. pH = 3.64
C. pH = .27
D. pH = 13.26
E. pH = 4.76
F. bases = ammonia, baking soda, soap, drano
Acids = vinegar, lemon juice, soda, fruit juices, milk
G. PH = 7 pOH = 7 note pH of 7 is neutral
$H^+ = 1.0 \times 10^{-7}$
H. pH of 0-6.9
I. Refer to glossary

Exercise 19-1

1. refer to text and glossary
2. refer to chapter 19
3. 238 ---------------------234
   92 90 alpha
   238 ---------------------238
   92 93 beta
4. .007 grams

20-1

1. Oxygen = 2 6 $1s^2 2s^2 2p^4$
   K L
   Calcium = 2 8 8 2 $1s^2 2s^2 2p^6 3s^2 3p^6 4s^2$
   K L M N
   Chlorine = 2 8 7 $1s^2 2s^2 2p^6 3s^2 3p^5$
   K L M
   Copper +2 = 2 8 17 2 $1s^2 2s^2 2p^6 3s^2 3p^6 3d^9 4s^2$
   K L M N
   Zinc = 2 8 18 2 $1s^2 2s^2 2p^6 3s^2 3p^6 3d^{10} 4s^2$
   K L M N
   Hydrogen = 1 $1s^1$
   K
   Mg = 2 8 2 $1s^2 2s^2 2p^6 3s^2$
   K L M
2. CH_4 = sp^3 Hybrid
   NH_3 = sp^3 Hybrid
3. refer to text
4. sp2 = 120°
   sp3 = 109.5°
5. K-Br
6. H:H
7. group 2
8. CsF is electronegative because of great difference in polarity.

Ideal Gas Laws Problem Set 1

1. 6.9 atm
2. 185 moles
3. 1.05 x 10^4 moles of SO_2
4. 730°K , -200°C
5. 3.83 atm yes an explosion occurs
6. 18.13ft^3
7. 3.47 atmospheres
8. 76cm = 1 atm
5. 6 liters = 0.25 moles
   484°R = 25°F

Gaw Laws Problem Set 2

1. H_2 is 2.83:1 faster than CH_4
2. The heavier gas weighs 43.52 grams.
3. 21.4 liters
4. 3.54 grams

Energy Key

1. Please use chemistry reference books and Easy Chem glossary.
2. Please use a chemical glossary, general references and Easy Chem glossary.
3. 4.286 x 10^{12} hertz
4. 1.09 x 10^{12} hertz
5. 3.83 x 10^{-20} joule/ hertz
6. 3.84x10^1m/s

Physical Properties of Solutions Problems

1. Refer to your Easy Chem glossary
2. 26.26mmHg
3. 31mmHg (vapor pressure at 30°C is 31.51mmHg)_
4. 22.72 grams
5. 1.127 x 10^5 g/mole

Colligative Properties

1. 345.42 grams
2. 431.75 grams
3. 179.3 grams

Rates of Reaction

1. Refer to EASY Chemistry notes and glossary.
2. 1.0769×10^{-4} mole HBr/liter mole
3. 1.04×10^{-4} xy/1sec

Metric System Quiz Key

a) 150°C = 302°F
   60°C = 140°F
   485°F = 251.66°C = 524.6°K
   1.75 liters = 1750mL
   375°A = 102°C
   195°C = 468°K
   160 mm = 16cm = 0.16m
   30 decimeters = 3 meters

b) 0.1875 g/cm^3

e) 2160 grams

d) 24 grams

Percent Composition Key

1. Au = 16.8 grams
   Ag = 4.76 grams
   Al = 2.8 grams
   Cu = 3.64 grams
2. 25% Au
   29% Cu
   19.78 grams of Ag
   46% Ag
3. 40% Carbon
4. 32.65% sulfur

Gas Laws Quiz Key

1. 1.96g/dm^3
2. 15 grams
3. 506.66mL
4. 249mL
5. 480cm^3 of Hydrogen, 16.25% N and 192cm^3 of carbon
6. 155mL
7.

moles	liters	molecules	grams
A) 0.41	9.22	2.46×10^{23}	14
B) 0.75	16.81	4.5×10^{23}	25.5
C) 0.6	13.44	3.6×10^{23}	20.4
D) 2	44.81	1.2×10^{24}	68

8. 810ml, 0.0357 moles N_2 and 2.143×10^{22} molecules of N_2
9. H_2S is 1.37 times faster than SO_2
10. a. decrease pressure
    b. increase pressure

Thermo Chem Quiz Key

1. 2800 calories
   2.8Kcal
   11704 joules
   11.7 KJ
2. 421.57 calories
   1762.18 joules or 1.762KJ
3. 30 grams of ice
4. 24543 calories or 24.5 Kcal
   102589.7 joules
   102.6 KJ
5. 0.6653 cal/g°c

Solutions Quiz Key

2.78 joules/g°c

1. 2.88 molar
2. 14.4 grams
3. 2 molar 240 grams
4. 1 molal 101.04°C
5. a. 33.33 grams b. 144 grams, 56.25 molar c. 2.5M

Mathematics of Chemical Equations Quiz

1. 8 moles of oxygen
2. 10 moles
3. 160 grams
4. 15 liters
5. 45.36 liters
6. 1.6 moles
7. 4 liters
8. 0.4 moles

Moles Quiz Key

1. 1.66×10^{-17} atoms
   1.88×10^{-15} grams
2. 2.22×10^{21} atoms
3. 5.5×10^{14} pennies
   9.166×10^{-10} moles
4. 0.003611 moles
   2.166×10^{21} molecules
5. 2 acres
   6. 0.0769 moles 4.614×10^{22}
   6.6×10^{-24} moles 4.33×10^{-22} grams

Periodic Table of the Elements

1 IA 1A	2 IIA 2A	3 IIIB 3B	4 IVB 4B	5 VB 5B	6 VIB 6B	7 VIIB 7B	8	9 VIII 8	10	11 IB 1B	12 IIB 2B	13 IIIA 3A	14 IVA 4A	15 VA 5A	16 VIA 6A	17 VIIA 7A	18 VIIIA 8A
1 **H** Hydrogen 1.008																	2 **He** Helium 4.003
3 **Li** Lithium 6.941	4 **Be** Beryllium 9.012											5 **B** Boron 10.811	6 **C** Carbon 12.011	7 **N** Nitrogen 14.007	8 **O** Oxygen 15.999	9 **F** Fluorine 18.998	10 **Ne** Neon 20.180
11 **Na** Sodium 22.990	12 **Mg** Magnesium 24.305											13 **Al** Aluminum 26.982	14 **Si** Silicon 28.086	15 **P** Phosphorus 30.974	16 **S** Sulfur 32.066	17 **Cl** Chlorine 35.453	18 **Ar** Argon 39.948
19 **K** Potassium 39.098	20 **Ca** Calcium 40.078	21 **Sc** Scandium 44.956	22 **Ti** Titanium 47.867	23 **V** Vanadium 50.942	24 **Cr** Chromium 51.996	25 **Mn** Manganese 54.938	26 **Fe** Iron 55.845	27 **Co** Cobalt 58.933	28 **Ni** Nickel 58.693	29 **Cu** Copper 63.546	30 **Zn** Zinc 65.38	31 **Ga** Gallium 69.723	32 **Ge** Germanium 72.631	33 **As** Arsenic 74.922	34 **Se** Selenium 78.971	35 **Br** Bromine 79.904	36 **Kr** Krypton 83.798
37 **Rb** Rubidium 85.468	38 **Sr** Strontium 87.62	39 **Y** Yttrium 88.906	40 **Zr** Zirconium 91.224	41 **Nb** Niobium 92.906	42 **Mo** Molybdenum 95.95	43 **Tc** Technetium 98.907	44 **Ru** Ruthenium 101.07	45 **Rh** Rhodium 102.906	46 **Pd** Palladium 106.42	47 **Ag** Silver 107.868	48 **Cd** Cadmium 112.414	49 **In** Indium 114.818	50 **Sn** Tin 118.711	51 **Sb** Antimony 121.760	52 **Te** Tellurium 127.6	53 **I** Iodine 126.904	54 **Xe** Xenon 131.294
55 **Cs** Cesium 132.905	56 **Ba** Barium 137.328	57-71	72 **Hf** Hafnium 178.49	73 **Ta** Tantalum 180.948	74 **W** Tungsten 183.84	75 **Re** Rhenium 186.207	76 **Os** Osmium 190.23	77 **Ir** Iridium 192.217	78 **Pt** Platinum 195.085	79 **Au** Gold 196.967	80 **Hg** Mercury 200.592	81 **Tl** Thallium 204.383	82 **Pb** Lead 207.2	83 **Bi** Bismuth 208.980	84 **Po** Polonium [208.982]	85 **At** Astatine 209.987	86 **Rn** Radon 222.018
87 **Fr** Francium 223.020	88 **Ra** Radium 226.025	89-103	104 **Rf** Rutherfordium [261]	105 **Db** Dubnium [262]	106 **Sg** Seaborgium [266]	107 **Bh** Bohrium [264]	108 **Hs** Hassium [269]	109 **Mt** Meitnerium [278]	110 **Ds** Darmstadtium [281]	111 **Rg** Roentgenium [280]	112 **Cn** Copernicium [285]	113 **Nh** Nihonium [286]	114 **Fl** Flerovium [289]	115 **Mc** Moscovium [289]	116 **Lv** Livermorium [293]	117 **Ts** Tennessine [294]	118 **Og** Oganesson [294]

Lanthanide Series	57 **La** Lanthanum 138.905	58 **Ce** Cerium 140.116	59 **Pr** Praseodymium 140.908	60 **Nd** Neodymium 144.243	61 **Pm** Promethium 144.913	62 **Sm** Samarium 150.36	63 **Eu** Europium 151.964	64 **Gd** Gadolinium 157.25	65 **Tb** Terbium 158.925	66 **Dy** Dysprosium 162.500	67 **Ho** Holmium 164.930	68 **Er** Erbium 167.259	69 **Tm** Thulium 168.934	70 **Yb** Ytterbium 173.055	71 **Lu** Lutetium 174.967
Actinide Series	89 **Ac** Actinium 227.028	90 **Th** Thorium 232.038	91 **Pa** Protactinium 231.036	92 **U** Uranium 238.029	93 **Np** Neptunium 237.048	94 **Pu** Plutonium 244.064	95 **Am** Americium 243.061	96 **Cm** Curium 247.070	97 **Bk** Berkelium 247.070	98 **Cf** Californium 251.080	99 **Es** Einsteinium [254]	100 **Fm** Fermium 257.095	101 **Md** Mendelevium 258.1	102 **No** Nobelium 259.101	103 **Lr** Lawrencium [262]

www.ingramcontent.com/pod-product-compliance
Ingram Content Group UK Ltd.
Pitfield, Milton Keynes, MK11 3LW, UK
UKHW051206260726
13967UKWH00011B/3143

9 781967 903658